东南大学建筑学院国际联合教学丛书
International Joint Teaching Series of SEU-ARCH

Typological Permanencies and Urban Permutations
Design Studio of Re-generation in Hehuatang Area, Nanjing

类型的恒在与城市的蜕变
南京城南荷花塘地块及住区建筑更新设计

鲍莉　Marco Trisciuoglio　姜蕾　湛洋　等著

江苏高校品牌专业建设工程资助项目（TAPP）
北京未来城市设计高精尖创新中心：城市设计
理论方法体系研究（UDC2016010100）

东南大学出版社
Southeast University Press

总 序

王建国
中国工程院院士
全国高等院校建筑学教育专业指导委员会主任
东南大学建筑学院前院长，教授，博士生导师

以践行"国际化"办学为宗旨的国际化教学正成为中国许多建筑院校新近最重要的设计教学趋势，这一趋势不仅包括了以往办学条件较好的诸如"老八校"和"新四军"之类的建筑院校，而且也覆盖到其他一些建筑院校。在我校举办的"2014年中国建筑院校境外交流学生作业展"中，参评学校已达35所。

从历史的角度看，国际化办学一直是世界名校建筑学办学的主要方式之一。美国哈佛大学、麻省理工学院和瑞士苏黎世联邦工业大学的建筑教育一直奉行的是全球延聘教授组织教学。中国早年的建筑教育也是一开始就与西方发达国家的建筑教学接轨。以中国建筑学办学历史最早的东南大学为例，当年的教材大纲、教程组织、教案编写等主要出自西方留学回来的教师之手，这些教师包括刘福泰、卢树森、鲍鼎、刘敦桢、杨廷宝、童寯、李汝骅、谭垣等，他们全都有着西方建筑学学习的经历。后来改革开放初期，也开展过一些外教教学的尝试，我本人就参加了由苏黎世联邦工业大学温克尔教授夫妇主持的建筑构成设计教学。而1946年创立的清华大学建筑系的奠基者梁思成也是美国学习背景。早期建筑设计教学的主要师资力量来自西方建筑教育的培养，所以在这个意义上讲，建筑设计教学的国际化也并不是全新的创举。只是在1950年代到1990年代，总体来说，改革开放前的中国建筑教育基本处于自我循环和封闭的状态。

General Preface

Professor Jianguo WANG
Academician of Chinese Academy of Engineering
Chair of National Supervision Board of Architectural Education, China
Former Dean, School of Architecture, Southeast University

Following the principle of "internationalizing" in architecture education, international joint teaching has recently become a trend among many architecture schools in China. Not only the schools ranked among "The Senior Eight" with more advanced education qualities, but also "The Junior Four", along with some other architecture schools, are chasing up with this trend. The number of schools participated in "Exhibition of Chinese Architecture Students Works of International Exchange, 2014", held in School of Architecture, Southeast University, already reached 35.

From a historic point of view, internationalized teaching is an ordinary phenomenon on among world-known architecture schools such as Harvard, MIT and ETHZ. These schools have always been employing professors worldwide for teaching. Actually, architecture education in China was in line with schools in western countries back in early years, taking the example of Southeast University, the architecture school with the longest history in China, the teaching program and curriculum of which at that time were all produced and organized by teachers graduating from western architecture schools, including Futai LIU, Shusen LU, Ding BAO, Dunzhen LIU, Tingbao YANG, Jun TONG, Ruhua LI and Yuan TAN, all of whom had overseas studying experience. Later in the early period of "Reform and Opening-up", we had also invited foreign teachers to our school as a teaching experiment. I, myself, had the experience participating in Architecture Composition Teaching Studio hosted by Prof. Heidi and Peter Wenger from Switzerland in 1983. Sicheng LIANG, who established School of Architecture, Tsinghua University also had studying experience in the U.S. Teachers involved in architecture teaching were mainly educated in the west in early years, therefore, internationalized teaching is not an innovation. Exceptionally, from 1950s to 1990s, China was basically self-circulating and enclosed.

今天，进入新千年的东南大学建筑教育走到了一个历史发展的转折点上，作为全国建筑教学的标杆，东南大学建筑教学必须应对当今全球建筑学领域学术的研究前沿和关注热点的流变。因此，我们将国际化作为新时期建筑教育努力突破的重点，而其中一个重要标志就是突破了以往多半教师先行出国学习进修，然后回校借鉴国际经验开展实验教学的做法，陆续开展了由境外教授和国际学生一起参与的、工作语言为英文的联合教学或工作坊项目。

多年来，东南大学建筑学院分别与美国麻省理工学院、加州大学伯克利分校、华盛顿大学、明尼苏达大学、德州农机学院和伍德布瑞大学以及瑞士苏黎世联邦工业大学、加拿大多伦多大学、荷兰代尔夫特理工学院、澳大利亚新南威尔士大学、奥地利维也纳理工大学和新加坡国立大学等合作组织了国际联合教学并取得显著成果。

经过多年的实践和持续积累，我们积累了较为成熟的国际合作办学和联合教学的经验。目前，东南大学建筑学院每年均开展 6～8 次国际联合课程教学，与国际知名建筑院系实现了校际学分互认，双授学位工作也在进行中。同时，东南大学建筑学院已经具备国际公认的办学特色和人才培养水准，拥有稳定和富有实效的国际联合培养的合作渠道，每年有一定数量的本科毕业生和研究生到国际知名建筑院系和规划设计机构继续深造和工作。

Stepping into the new millennium, Southeast University is coming to a turning point. As a role model among all architecture schools in China, it needs to deal with the changes of academic frontiers and heat focuses of architecture education in a global scale. As a significant symbol, Southeast University steps beyond the tradition of sending teachers abroad for learning, who returns with foreign experience in order to conduct experimental teaching practice back home, rather, conducts series of international joint teaching programs or workshops involving foreign professors and international students, taking English as the official working language.

For years, our school has respectively carried out international joint teaching with MIT, UC Berkeley, University of Washington, University of Minnesota, Texas A&M University, Woodbury University, ETHZ, University of Toronto, TU Delft, University of New South Wales, TU Vienna and National University of Singapore, with significant outcomes.

After many years of practice and experience accumulation, we have gained mature experience in internationally collaborative education and joint teaching. So far, on average, our school carries out 6 to 8 joint teaching courses every year. We are also working on mutually recognizing credits and double degrees recognition together with world-known architecture schools. In the meantime, our school is internationally recognized by its education characteristics and qualities, having built up stable and substantial connections with world-known architecture schools for international joint education. Each year, a considerable number of undergraduate and graduate students continue their study in these schools or begin working in international institutions worldwide.

在教学实践中，我们也曾克服了不少实际的困难，如国内外学校的学期时段设置和教学计划安排存在的差异，不同文化背景的师生在合作交流时存在的价值观差异，以及教学经费筹措、教学活动管理、教学空间安排乃至师生的安全保险等等。为使这项教学活动实际可行，我们是从研究生阶段启动国际联合教学点，主要是教学时间和计划较为灵活，便于组织安排，同时研究生各方面较为成熟，境内外自主交流沟通和生活自理比较有保障。随着不断发展和经验积累，目前国际联合教学已经扩大到本科教学，课程设置和选题也有部分已经直接遴选在国外基地，让学生学习国外场地调研工作和人际沟通的能力。一直以来，东南大学建筑学院对国际联合教学工作中的教师人员配备给予优先，并在学院层面划拨了专门的经费加以支持。

国际联合教学极大地激发了学生的学习热情，使他们有机会直面国际化的教学授课环境，感受不同的教学传统、文化特点和创新活力，显著开拓了国际化视野，使学生们在日后深造和就业竞争中直接受益。而同时，国际联合教学对于教师也是一次参与和感受国际化教学环境的极好机会。

多年来，我们切身感到国际联合教学对彰显东南大学建筑教育特色和优势的益处，此举也是东南大学建筑教育和办学国际化的重要组成部分。本丛书的陆续出版，从一个侧面见证了上述国际化教学方面取得的成果，由于很多工作仍然属于探索性的尝试，所以难免缺憾，希望读者批评指正。

We have also overcome many hurdles during the practice of joint teaching, such as the differentiations of academic schedules and curriculum between different schools, different perceptions of values of teachers and students from various cultural backgrounds, financing issues, management of teaching activities and teaching space, and even security insurance for each individual involved. To make these joint teaching programs more promising, at the beginning, we activated the teaching points only in graduate students' curriculum, since they have more flexible schedules for teaching organization, and they are generally more mature in every aspect, so that they can autonomously communicate with others and organize themselves. Along with the continuous development and experience accumulation, international teaching program has been extended to undergraduate curriculum. Some sites located abroad are chosen to offer better opportunities for students to learn the skills of site analysis and communication in foreign contexts. Our school has long been giving priorities to international joint teaching program, providing advanced teaching resources and extra financial support from school.

International joint teaching program greatly encourages students' enthusiasm of studying, offering them opportunities of facing directly to internationalized teaching environment, coming across with various teaching traditions, cultural characteristics, and creative energies, broadening their perspectives internationally, for their own benefits in future studies and careers. At the same time, International joint teaching program is also a perfect chance for teachers to participate into international academic community.

Over years, we have experienced the benefits brought by international joint teaching, which enhances Southeast University's characteristics and advantages in architecture education. International joint teaching forms an important part of architecture education in Southeast University. Publication of this series of books, from a side aspect, witnesses the outcomes of international joint teaching mentioned above. Since a lot of work is still under experimental practice, some regrets are hard to avoid, any correction and comment are sincerely welcome.

目 录
CONTENTS

前言：身份的模糊性　　　　　　　　　　　　　　　009
Preface: The Ambiguity of Identity

01 背景：荷花塘的城市形态　　　　　　　　　　　017
Background: Hehuatang-urban Form

02 教学设计　　　　　　　　　　　　　　　　　　027
Teaching Proposal

03 教案框架　　　　　　　　　　　　　　　　　　035
Program Framework

04 前期研究　　　　　　　　　　　　　　　　　　039
Preliminary Study

05 设计成果　　　　　　　　　　　　　　　　　　045
Design Works

06 写作训练　　　　　　　　　　　　　　　　　　155
Writing

07 师生反馈　　　　　　　　　　　　　　　　　　187
Reflection

后记：基于城市形态类型学的当代设计教学实验　　　193
Epilogue: A Contemporary Design Teaching
Experiment Based on Urban Morphology Typology

前言
Preface

身份的模糊性
马可·德诺西欧

The Ambiguity of Identity
Marco TRISCIUOGLIO

在对南京荷花塘地区的方案设计中，我们与东南大学的本科生及硕士生一起，针对其身份的模糊性进行了探讨：今天，在定义相对清晰的历史城区的城市更新工作中，生成的新建筑应该有什么样的特征？

1963年，奇安弗兰科·卡尼吉亚在他关于科莫[1]的著作中用这样的话来开头："认识城市结构是进一步操作城市的准备条件。"而这一研究的目标就是绘制意大利科莫的类型学地图。类型学地图是由确定时期城市聚居地所有建筑物地面层平面图组成的地图。其绘制从现存地图（在欧洲通常是19世纪的地籍图）开始，要借助大量的本地调研活动，而最重要的是需要引入基于类型学与类比思想的城市形态学的观察方式。

卡尼吉亚将存在过的一系列城市——过去各个时代的科莫，作为规划和设计新城市——未来时代的科莫的前提。 不同历史图层的地层学式的叠图可以为城市自身的未来发展暗示各样的轨迹、线条和轮廓。

以城市空间形态学的方法作为设计活动的基础，在奇安弗兰科·卡尼吉亚的科莫研究50多年后的今天，仍然在发挥作用。

例如，薛杰，在他极富影响力的著作，在"中国生态城市与未来可持续建筑"合作项目框架下发表的《城市与形态：关于可持续城市化的研究》（2011）一书中对形态学方法进行了重新阐述（甚至加入了建设城市及环境的可持续发展问题）[2]。几乎在同一时间，陈飞和凯文·恩韦茨在利物浦和谢菲尔德完成，并在伦敦（Ashgate出版社）出版了他们的《中国城市设计：类型形态学方法》（2013）。在这一工作中，卡吉尼亚式的理论和方法发挥了基础性的作用。

We worked, in our projects for the small house settlement of Hehuatang in Nanjing, together with the Undergraduate and the Master students at the SEU School of Architecture, on the ambiguity of identity: which can be the character of contemporary architectures in approaching the urban regeneration of a well defined historical settlement in nowadays China?

In 1963, Gianfranco CANIGGIA, opens his book on Como[1] using these words: "Recognizing urban structures is a condition to operate on cities". The goal of that research is drawing the typological map of the Italian village of Como. The typological map is a map of the first floors of all the buildings of an urban settlement in a certain date, drawn starting from the existing maps (usually in Europe the cadastral maps of 19 century), using a local intense survey's activity, but overall adopting an urban morphology reading method based upon the ideas of typology and analogy.

CANIGGIA keeps the existing cities (the lot of "Como" in the past times) as a premise to plan and design the new city (the Como of the future times). The stratigraphic overlapping of historical layers can suggest traces, lines, shapes for the future development of the city itself.

The morphological approach to the urban space as a basis for designing activities is still working nowadays (more than fifty years after the researches on Como by Gianfranco CANIGGIA).

Serge SALAT, for example, renewed the morphological approach (even adding the sense of sustainability in building cities and their environment) in his monumental book *Cities and Forms: On sustainable Urbanism* (2011), published in the framework of the cooperation project "Chinese Eco-cities and Sustainable Buildings in the Future"[2]. Almost at the same time, Fei CHEN and Kevin THWAITES wrote between Liverpool and Sheffield and published in London (through Ashgate publishing company) their *Chinese Urban Design: The Typomorphological Approach* (2013). That is a work where Caniggia's theories and methods play a fundamental role.

由于英国地理学家怀特汉德和新西兰的规划学者谷凯的努力，关于中国城市居住地的形态学式研究方法（与遗产保护问题相联系）在过去十年中已经十分活跃[3]。但是，康泽恩的概念更多地基于各种空间和使用的等级结构，而不是城市的建造和建筑物的现实状态（它们的类型、风格元素和建筑材料）。

也许改善中国城市的城市形态学研究，引入卡吉尼亚和穆勒托里的方法，在中国城市设计的新趋向下可能会是成果丰硕的。而如果这件事直到现在还没有发生，就像邓浩最近提及的，可能与英语或者中文翻译资料的缺乏有关："与康泽恩学派相比，意大利学派的工作在中国依然鲜为人知。这在很大程度上要归因于这样的事实，就是穆勒托里学派的工作几乎没有可查考的英文版，更鲜有中文版本。"[4]

所以我们现在面临着双重的挑战：利用意大利的类型形态学方法（与萨韦里奥·穆勒托里和奇安弗兰科·卡吉尼亚的研究相联系）来调研中国城市，通过相关建筑和空间的设计找到一种新的操作城市的方式，这同时也将富有案例和实验的新生命带回给类型学研究本身，它在过去30年的意大利建筑学论战中实际上并非如此幸运。我们有机会面对这一双重挑战，皆从类型形态学在中国城市现实和城市设计中的创新性的应用而起，它与那些在意大利研究刚刚兴起时的意大利本土实践是非常不同的。

虽然与意大利城市完全不同，中国的城市却正在经历着与1960年代意大利的历史城市同样的情况，也正是此时阿尔多·罗西写成了他的《城市建筑学》[5]：城市涉及新城市社会和经济驱动力的方面吸引了很多的讨论，而全景式的当代中国城市本身的形态及与此相关的建筑类型和建筑特征却少有提及。

Because of the efforts by the British geographer J.W.R. WHITEHAND and the New Zealander planner Kai GU, the morphological approach to the Chinese urban settlement (linked with the question of the heritage's conservation) was already vivid during the last decade[3]. But the CONZEN's conception is more based upon the hierarchy of spaces and of uses, rather than upon the construction of the city and the reality of buildings (their types, the stylistic elements and the building's materials).

Maybe an improvement of the urban morphological studies on the Chinese city, adopting the CANIGGIA's and MURATORI's methods can be really fruitful for the new tendencies of urban design in China. If this didn't happen until now, as Hao DENG recently highlighted, it is because of a lack of an English (or Chinese) translation: "the work of [the Italian] school remains less known in China than that of the Conzenian school. This is owing to in substantial part to the fact that much of the work of the Muratorian school is still not available in English and even less of it is available in Chinese."[4]

So, we are living a double challenge nowadays: using the Italian typo–morphological approach (the one linked with the studies by Saverio MURATORI and Gianfranco CANIGGIA) to investigate the Chinese cities and to find a new way to operate in them by designing their buildings and their spaces, and giving a new fruitful life of experiments and cases to those typological studies, at the end non to so lucky in the Italian architectural debate of last 30 years. We can face this double challenge just thanks to the innovative application of the typo–morphological approach to the Chinese urban realities and urban design, that are so different from the Italian ones that were considered at the rising of those researches in Italy.

Though it is totally different from the Italian city, in the Chinese city of nowadays is happening exactly what in the historical Italian cities happened in the Sixties (when Aldo ROSSI wrote its *The Architecture of the City*[5]): there is a great debate around them, concerning the new urban societies and their economical dynamics, but few things has been told in the last years about a panoramic or a comprehensive review of the shape of the contemporary Chinese cities themselves and the architecture and the buildings' characters related with that .

不幸的是，在世界范围内关于城市的讨论中，爱德华·格莱斯的书，《城市的胜利——城市如何让我们变得更加富有、智慧、绿色、健康和幸福》（2011），以及乔纳森·F.P. 罗斯的《顺服的城市——关于城市生活的未来，现代科学，古代文明和人性告诉我们什么》（2016），似乎是简·雅各布斯的《美国大城市的死与生》（1961）的续集，甚至在内容中都是：如此普遍，如此宽泛，如此基于北美的（既不是欧洲的，也不是亚洲的）城市经验。所有这些作品都可以叙述当代城市，但它们却无法向设计师提供实用的建议。

这就是为什么我们决定重新开始，跟我们的学生一起，通过绘制地图、调研、勾画建筑平面草图、拍照理解城市空间——集体和私人空间及其使用，来进行城市分析。

荷花塘地区位于南京历史城区的南部（城墙范围之内），是一个重要的实验场地：在这个有许多棚屋、破败房屋、狭窄街道，日常生活被封闭在南侧紧邻的高大城墙之下的"贫民窟"中，我们可以识别出在该地区反复出现的一些"类型"。一些历史悠久的院落住宅为其他部分留下可追溯的足迹：一个通常很小的庭院是一个多家庭建筑群的真正核心；有时内部非常狭窄的街道可以穿过两三个以上的院落，连接两条平行的道路；同一条路上的房屋"使用"那条道路的方式是同时作为私人和集体空间。这些都是需要被调查并检查其形态和成因的主题。

这种推理推动了我们开始描绘荷花塘的类型地图：它是这个场地的局部地图，但它远比标识建筑保护等级（标记为"保护建筑""历史建筑""普通建筑"）的地图更能说明这个小地区的情况。

Unfortunately, in the worldwide debate about cities the books by Edward GLAESER, *Triumph of the City: How Our Greatest Invention Makes Us Richer, Smarter, Greener, Healthier, and Happier* (2011), and by Jonathan F.P. ROSE, *The Well-Tempered City: What Modern Science, Ancient Civilizations, and Human Nature Teach Us about the Future of Urban Life* (2016), seem to be the sequel of Jane Jacobs' *The Death and Life of Great American Cities* (1961), even in the contents: so generic, so wide, so based on the experience of the North American city (that is neither the European one, nor the Asian one). All those works can describe narratively the contemporary cities, but they are unable in giving practical suggestions to designers.

That's why we decided to start again, with our students, from an urban analysis made by mapping, surveying, taking sketches of buildings' planes, taking pictures to understand the urban spaces, the collective and the individual spaces and their use.

The Hehuatang area, in the south part of historical Nanjing (inside the precinct of the City Walls), is an important experiment's field: in the "slum" of a lot of shacks, poor houses, narrow streets, a daily-life closed against the incumbent height of the south city walls, we could recognize some "types" that recur in the area. Some historical courtyard-houses give the footprints for others: a (often very) small courtyard is the real core of a multiple homes' buildings; internal very narrow streets can cross more than two or three courtyards to link two parallel roads; the way in which the houses along the same road are "using" that road, as individual and collective spaces at the same time. They are all topics to investigate and to check in its morphologies and in its interpretations.

This kind of reasoning pushed us in starting tracing a typological map of Hehuatang: it is a partial map of the site, but it tells about that small district much more than the map devoted to introduce the level of protection of buildings (labelling them as "protected buildings", "historical buildings", "ordinary buildings").

使用类型来描述城市意味着进入房屋和城市空间的遗传结构，意味着识别那些迄今为止建造城市所依据的法律和规则，意味着理解在这一地区的发展中新方案、新设计活动、新经济与社会项目继续前进的形式基础。

方案，事实上，也理所当然，是此次工作的真正重点。

通过形态类型学方法介入荷花塘，意味着提出"身份"的问题。借助这一概念，我们考虑了当下可以设计的新建筑与类似荷花塘这样地区的过去建筑传统之间的关系。在中国城市的"城市遗产性居住区"中，什么样的建筑特点是更合宜的？

这不是一个简单的问题，而是当今中国城市和中国建筑新趋向的关键问题。我们确信荷花塘新建筑的风格不可能像多数现代亚洲城市，从杜拜到上海[6]，一样的"新国际风格"。另一方面，中国社会的富裕阶层正把越来越广泛传播的"新传统中国建筑"风格视为理所当然。这种风格在整个中国都高度相似（也许与西方理论中的"新城市主义"相关），而对于它们被建造地区的真实特征却兴致寥寥，因而带有一点"流行"与"附庸风雅"的意味。虽然老门东已然如此，现在荷花塘的新规划方案也有类似的建筑风格，我们却认为这不应该是荷花塘这样的历史住区的唯一选择。

就在过去的十年中，中国新一代建筑师的研究，一些更加有趣而丰富的东西正在被唤起。在新国际风格和新传统风格之间存在着第三种方式。这一方式研究设计新形式和新语言，同时具有强烈的创新性（例如对可持续性的研究），并深深扎根于中国建筑的真实传统。

Using types to describe the city means entering inside the genetic structure of houses and urban spaces, means recognizing the laws and the rules according to which the city has been built until now, means understanding the formal basis to go ahead in the development of that site with new projects, new design activity, new economical and social programs.

Projects are in fact and of course, the real focus of this work.

Approaching Hehuatang through the typo–morphological method means putting the question of "identity". Using this concept, we consider the relationship between the new architecture we can design nowadays and the tradition of past architecture in a place like Hehuatang. Which is the more appropriate character of the architecture in the urban heritage settlements in the Chinese city?

This is not a simple question, but it is a crucial question for the Chinese cities of nowadays and also for the new tendencies of Chinese architecture. We believe that the character of the new architecture in Hehuatang cannot be the neo-international style of most of the contemporary Asian architecture (from Dubai to Shanghai[6]). At the same time the richer classes in the Chinese society are feeling as a "must" the more and more spread taste for a "neo-traditional Chinese architecture". This taste is always the same all over China (maybe linked to the western philosophy of New Urbanism), with very few interest for the real features of the place in which they are built, maybe a little "pop" and a little "snob" at the same time. Even if the near district of Laomendong is already pervading now also Hehuatang with the same architectural style, we can keep that as not the only possibility for the renewal of a historical settlement like Hehuatang.

The research of the new generation of architects in China is evoking, just in the last decade, something more interesting and more fruitful. There is a third way in between the neo-international style and the neo-traditional style. It is the way of researching through design new shapes and new languages, at the same time strongly innovative (for instance in their research on sustainability) and deeply rooted in the real tradition of Chinese buildings.

我们需要考虑"传统"究竟是什么：在形式、类型化解决方案、建造方法的无尽传递中，每代人都加入了自己的部分来进行创新。张雷与张雷联合建筑事务所、华黎与迹·建筑事务所、大舍建筑设计事务所、朱锫、张轲与标准营造、刘家琨、直向建筑等，与享有国际声誉的创建了非常建筑和业余建筑工作室的大师们一起，都是一项重要变化的信号。我们决定朝这个方向工作，将中国设计的第三种路径与基于意大利穆勒托里和卡吉尼亚学派的城市分析相结合。

我们鼓励学生去发掘深入的院落住宅类型调研可以给新的、现代的方案带来何种契机，去尊重场地的本性。也就是所谓的"场所精神"，就是每个人都可以察觉到的那些。当人在荷花塘的狭窄街道上漫步时，在打量那些小院子时，在望进那些窗户后面的简陋餐厅时，在闻到那些在家门前烹制的食物的味道时，或者在听到从某些公共区域传来的无尽而懒散的麻将碰撞声时。

我们方法的真正秘密在于保持历史与我们之间的距离。我们希望以一种分离，甚至有时也带一点戏谑的方式来看待老的中国建筑以及类型和模式。我们认为"我们的"荷花塘设计逃离了身份的模糊性，是一个"旧住区的新编"，就如同1935年鲁迅的书中对那些故事的"新编"。

最近，在介绍一个纽约所罗门古根海姆博物馆的中国当代艺术的艺术展览时，翁笑雨写道："旧传说不仅是传说，也是组成历史本身的活动与事件。鲁迅不是简单地重写童话故事，而是开放了过去本身以进行重新思考"。[7]

"重新思考过去"，这似乎是设计"新的"中国历史城市的关键词，也是牵涉着我们所有人的一项使命。

We should consider "tradition" what really is: the endless passing from hand to hand of shapes, typological solutions, building methods, in which each generation gives its support of innovation. The architectures by Lei ZHANG and AZL Architects, Li HUA and TAO, Atelier DesHaus, Pei ZHU, Ke ZHANG and ZAO/Standard Architecture, Jiakun LIU, Vectror Archtects, among the others, are all signals of an important change, together with the internationally recognized Chinese great masters, the ones who founded FCJZ and Amateur Architecture Studio. We decided to work in that direction, connecting this third way of Chinese design with the urban analysis based on the Italian school of MURATORI and CANIGGIA.

We encouraged our students to investigate the opportunities given by a deep study on the courtyard-house type for new and modern projects, able to respect the sense of the place. It is the so called "genius loci", that everybody could perceive walking in the narrow streets of Hehuatang, looking into the small courtyards, watching inside the windows in those poor dining rooms, feeling the smell of foods often cooked in front of the home's door, listening to the sound of the mahjong's small tiles coming from some collective spaces, during endless and indolent matches.

The real secret of our approach is taking some distance between the sense of the history and ourselves. We would like to look at the old Chinese architectures and typologies and patterns with a good detachment, sometime also with a little bit of irony. We like thinking that "our" Hehuatang, designed in escaping from the ambiguity of identity, is an "old settlement retold" as how the tales were "re-told" in the book by Lu Xun in 1935.

Recently, introducing an art exhibition on Chinese contemporary art at Solomon Guggenheim Museum New York, Xiaoyu WENG wrote: "Old tales are not only legends, but also the very past events and incidents that constitute history as such. Rather than simply rewriting fairy tales, Xun LU was opening up the past itself for reconsideration".[7]

"Reconsideration of the past". This seems to be the keyword to design the "new" historical Chinese city and this is the mission we all are engaged about.

Notes

1. Gianfranco CANIGGIA. Lettura di una città: Como [Reading a city: Como]. Rome: Centro Studi di Storia Urbanistica, 1963.

2. It is a research project lead by the French CSTB (Urban Morphology Laboratory), directed by Serge Salat, and the Shanghai Research Institute of Buildings Sciences together with the College of Architecture and Urban Planning at Tongji University. Serge SALAT, Françoise LABBÉ, Caroline NOWACKI. Cities and Forms: On Sustainable Urbanism. Paris: Hermann Éditeurs des Sciences et des Arts, 2011.

3. J W R WHITEHAND, Kai GU. Urban conservation in China: Historical development, current practice and morphological approach. Town Planning Review, 2007, 78(5): 643-670.

4. Hao DENG. Rossi or Muratori? A Chinese perspective // Viewpoints: Discussion of topical issues in urban morphology. Urban Morphology, 2016(20).

5. Aldo ROSSI. L'architettura della città. Venezia: Marsilio, 1966.

6. The two examples are not by chance. They come from the book by Daniel BROOK. A History of Future Cities. New York/London: Norton, 2014, a research on the identity of Asian cities that have a native western footprint.

7. Xiaoyu WENG, Counter-mythologies, or Tales of Our Time, introduction to the catalogue (re-told). Tales of Our Time, exhibition organized by Xiaoyu WENG and Hanru Hou, Solomon Guggenheim Museum, New York, November 4th 2016 – March 10th 2017.

01

背景：荷花塘的城市形态
姜蕾

Background: Hehuatang-urban Form
Lei JIANG

2015年荷花塘历史文化街区保护规划

2015年公布的《荷花塘历史文化街区保护规划》[1]及2013年公布的公众意见征询版（图1），展示给我们的是一个典型的当代中国保护及更新规划的例子——致力于在老城区内"保护并延续具有典型传统（明清时期）民居特征的街区整体格局和风貌"，正如这一保护规划中所阐明的。类似的体系和做法也同样被应用在南京的其他历史地段，如老门东、南捕厅等，而众多其他历史城市情况也大致相似。

荷花塘整个地块大约126 000平方米。其中可以了解到，25%的建筑物建成于清朝以前；14%在中华民国时期建成；很大一部分，大约42%的建筑，在1950—1980年间建成；而剩余的19%可以视为当代建筑，1980年以后建成。在涉及建筑高度的数据中，可以了解到68%属于1层建筑，而2层建筑占25%。

这些数据突显了这一区域与周围区域的不连续性，其他区域大量被多层、高层现代建筑所覆盖。更新规划倾向于保持这些差异。原有聚落可以看做传统院落住宅演化产生的不同类型的组合与复合，这一城市肌理突显了传统住宅类型的高适应性。基于此种背景，像前文提到的，荷花塘的保护更新规划致力于"保护并延续具有典型传统民居特征的街区整体格局和风貌"以及引入旅游功能。

The Conservation Plan of Hehuatang, 2015

The Conservation Plan of Hehuatang, which was published in 2015[1] together with its public consulting version (Figure 1) in 2013, is showing us a typical contemporary Chinese conservation and regeneration urban design aiming to "protect and conserve the entire layout and features of the historical area with typical characteristics of traditional (Ming and Qing) houses" in a historical area, as that emphasized in the conservation plan. Actually, similar system and practices are also introduced and practiced in other historical areas around Nanjing, like Laomendong, Nanbuting, etc, and also in many other historical cities.

The total area of Hehuatang is more or less 126,000 sqm. Inside we can find that 25% of the buildings were built before Qing Dynasty; 14% were built during Republic of China; the majority, 42%, were built in 1950-1980; and the rest, 19% could be considered as contemporary buildings, after 1980. Another interesting data concerns the height of the constructions. In fact we can see that 68% are one-storey buildings and 25% have two stories.

All these highlights a discontinuity of this area compared with many surrounding areas, as most of them are composed by multi-story or high-rise contemporary buildings. The regeneration plan tends to preserve those differences. As the original settlement is a complex, a combination of different types derived from the evolution of the traditional courtyard houses, this urban tissue demonstrates the resilience of the traditional type of houses. With such a circumstance, as referred before, the conservation and regeneration plan of Hehuatang aims to "protect and conserve the entire layout and the features of this historical area with the typical characteristics of traditional houses", as well as introduce tourism into this area.

图1 荷花塘历史文化街区保护规划，2015年
来源：南京市规划局，获取自 http://www.njghj.gov.cn/

Figure 1　The conservation plan of Hehuatang, 2015
Source: Nanjing Urban Planning Bureau, retrieved from http://www.njghj.gov.cn/

通常，这类保护计划主要是在土地利用型城市规划的框架下进行开发的，而不同区域之间的关系则被切断。城市主要被视做一个"静态的实体"，关注点多在于建筑外部风貌，比如建筑物的特征是否同质，以及足以唤起人们对中国某段过往历史时期的"记忆"。如果现状并非如此，可以对其进行调整或改变以达成这一特定的期望或目标，令其高度同质，让人印象深刻。而真实的城市过程，以及随时间积累而成的地区内部或相互间的关联性则在这类规划中相对被忽视。

这种策略实际上可以追溯到甚至比北京1999—2004的实践(WHITEHAND, GU, WHITEHAND, et al, 2011：172)更早的年代，比如1984年前后南京夫子庙地区的重建项目，在改革开放的背景下，其主要目的是吸引游客与投资(XIE, COSTA, 1991：293)。然而，它们还应当继续作为这类"保护"的主要动机吗？我们继续以此种方式工作，是有意识的行动抑或惯性使然？一些重要的问题亟待重新思考。

必须承认，目前"一个综合的、可以为保护提供理论依据的关于城市景观发展史的知识体系"的建立过程是缓慢的(WHITEHAND, GU, WHITEHAND, et al., 2011：172)。而这里，借助荷花塘这一地区，我们希望尝试接近这一知识体系。

荷花塘地区城市形态与南京城南

依据南京的保护规划，荷花塘是南京老城九片历史区之一，位于城南，明城墙脚下。按照城市边缘带理论[2]，它属于连续的南京城市内边缘带的一部分。借助街道格局、土地利用格局以及形态区域的分析，南京这一连续内边缘带的基本构成及其形成过程可解析如下。

Usually, such conservation plan was mainly developed under the framework of the land-use urban planning, and in turn, the relation between different blocks is cut off. Then the city is mainly viewed as a "static physical entity" and most of the interest is on the external features, on whether the features of the buildings are homogeneous and seem able to recall the "memory" of a certain historical period of China. If not, they can be adjusted or correct to satisfy that special expectation or purpose, homogeneously and impressively. However, the real urban process and the relation inside or among areas accumulated in the course of years are comparatively neglected in such plans.

This kind of strategy actually can be traced back to the years even earlier than the practices of Beijing from 1999 to 2004(WHITEHAND, GU, WHITEHAND, et al, 2011: 172), to the years of the reconstruction practices of the Confucius Temple of Nanjing around 1984, which mainly aims to attract tourists and investments(XIE, COSTA, 1991: 293), under the influence of economic reforms. But should that still acts as the major intention or motivation for such "conservations"? We are working in this way with awareness or just with inertia? There are several important questions we should ask.

It must be admitted that "a coherent body of knowledge of the historical development of urban landscapes that might inform a theoretical basis for conservation" has been slow to develop(WHITEHAND, GU, WHITEHAND, et al, 2011: 172). And here, with Hehuatang area, it's an attempt to approach that system of urban knowledge.

Urban form of Hehuatang area and Nanjing South

According the conservation plan of Nanjing, Hehuatang is one of the 9 listed historical areas inside the old city of Nanjing. It's an area in city south, under the Ming City Wall. Referring to the urban fringe belt theory[2], it can be considered as part of the continuous inner fringe belt of Nanjing. Taking use of street pattern, land-use pattern, and the morphological region, the general composition of this continuous inner fringe of Nanjing and its formative process will be clarified as following.

以时间线考察

根据1929年的航空照片（图2），我们可以发现，在稠密的建成区之外，仍然有一些城墙附近的区域作为耕地或空地被遗留下来，基本在内部的两个角落附近或者分散在城墙外。还有两个在通商口岸时期(1842—1912)建成的工厂，一个是在西北方向（Ⅰ）的清代江南铸造银元制钱总局，成立于1897年，另一个是在城墙外东南方向（Ⅱ）的清代金陵机器制造局，成立于1865年。而在1929年直到1949年，城市发展过程很可能是以工业时代大尺度单元的填入为主的内边缘带的逐步填实（内蕴）过程。

至于1949年以后，历史过程的研究暗示，在社会主义早期，到1978年以前，甚至到1990年代，这里的建设以"单位"为主，如工厂或机构等，包括职工宿舍和相关设施。它们很多也可以被视为工业时代的大尺度单元。在老城西南角的前工业地块可以认为属于这一类型（Ⅲ）。另外城墙外西南部的一大块曾经是工厂和仓库用地的城市"休耕"区域面积（空地）（Ⅳ）也属于这一类型。实际上，在这个区域内的所有工厂或前工厂及其设施都在经历着替代、重建或改造。根据城市历史，可以知道，这与1990年代以后，工厂开始迁出市区，其建筑逐渐废弃有关 (LI, ZHOU, 1995a, 1995b)。也在那个时期前后，商品房封闭式社区开始出现，这与改革开放后土地使用权变化，如土地使用权和所有权的分离相关联[3]。另外，在同一时期，1978年以后，现代的基础设施和设备、集中的商业区、高层建筑等等也纷纷出现了。

A review with the timeline

Based on the 1929 aerial photo (Figure 2), we can find that beside the dense built-up area, there are still some areas left as arable land or open land near the city wall, generally near the two corners inside and scattered outside the city wall. There are also two factories, which were built during the Treaty Port Period (1842-1912). One in the northwest (I), which was Jiangnan General Mint of Silver Dollars in Qing dynasty, built in 1897. The other is in the southeast outside the city wall (II), which was Jinling Machinery Manufacturing Factory in Qing dynasty, built in 1865. Then during the years after 1929, until 1949, the development is likely to be a repletion process of the fringe belt with coarse-grained units of industrial era.

As to the years after 1949, the study of the history implies that, in the early years of the socialist period, the time before 1978, or even to 1990s, the constructions here would be mainly danwei, as factories or institutions, etc, with its accommodations and facilities. Many of them are also kinds of coarse-grained units of industrial era. The former industrial area at the southwest corner can be recognized as this type (III). Then a big piece of urban fallow area (open land) (IV) outside the city wall at southwest where used to be factories and warehouses, is also this type. Actually, all the factories or former factories with their facilities in this area generally are all under replacement, reconstruction or transformation. Considering the history, we know, that is associated with the fact that in 1990s and after, the factories started to move out of the city and the constructions of them were gradually abandoned(LI, ZHOU, 1995a, 1995b). Around that time commodity gated communities started to take their place, related with the changes of land ownership after economic reforms, like the separation of land usufruct right from the land ownership right[3]. Also in the similar time, some years after 1978, modern infrastructures and facilities, concentrated commercial areas, high-rises, etc, also arrived.

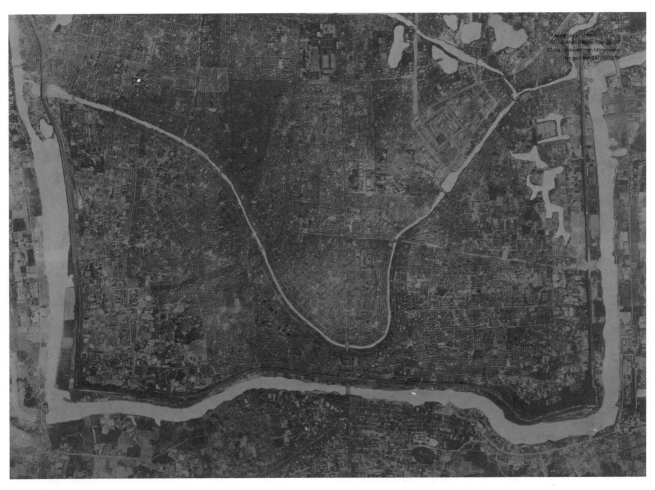

图2 1929年南京航空照片（局部）
来源：美国国会图书馆地理与地图部，获取自
https://www.loc.gov/item/2012593209/

Figure 2 Aerial photo of Nanjing in 1929 (detail)
Source: Library of Congress Geography and Map Division, retrieved from
https://www.loc.gov/item/2012593209/

到现在为止，除了一些城市"休耕"区，只有部分绿地被遗留在高密度的建成区之外。首先是沿着城墙（V）的一些绿地，它们与东北方向（VI）的一处公园（白鹭洲）连接。另外还有一处由传统的私家花园更新改造而建成的公园（愚园），在西南部（VII）。它们被遗留下来主要是由于特殊的地形以及自身历史价值（图3）。

现在，我们把上述这些特殊的，大多是大尺度的地块（图3，蓝、绿部分），如工厂（或前工厂）和公园提取出，进行进一步的讨论。对于这一区域的其他部分，其中包括荷花塘地区，我们将借助地块模式进行分析。

以地块模式考察

当我们开始再次读图，进行地块尺度和地块轮廓等方面的分析，可以进一步帮助完成按时间分层的工作。在1987年前后使用权（土地使用权）的分离使得城市建设和房地产开发项目更有可能在城市致密的建成区内取得较大尺度、较规整的地块以实现项目。而在这之前，建设通常以零散的方式进行，多为小尺度和不规则地块。据此，可以大致将区域内剩下的部分分为两组。一类是轮廓规则的大尺度地块（图3，橙色部分），第二类是那些含有轮廓不规则的小尺度地块的区域（图3，深灰部分）。

第一组更容易被理解，它们主要是新建住宅社区及商业地区和现代设施的组合，通常是多层或高层。而第二组似乎相对难以辨认。它们大多是对传统聚落和院落住宅进行零碎的更新替代的结果，其基本格局形成于土地使用权分离以前，也与社会主义改造过程相关联。这类地区通常混合了老宅和少量小规模工厂，多层住区也散布其中，而荷花塘就是这类地区的一个典型。

Until now, besides some urban fallow areas, only some green spaces are left behind by the high dense construction. First is some green area along the city wall (V), connecting with a park at northeast (VI). Then there is also a park based on some renovation and reconstruction of a traditional private garden, in the southwest (VII). They were left there mainly because of the special terrains together with their historical value (Figure 3).

Now we take all the above mentioned special, mostly coarse-grained plots (Figure 3, blue, green) of this area out, like factories (or former factories) and parks, to make some further discussion later. As to the rest parts of this area, including Hehuatang area, we will continue the analysis with the plot pattern.

A review with the plot pattern

Then when we start to read the map again, there are other aspects helpful to realize the stratification by time, the scale and the outline of the plots. The separation of the usufruct right (land-use right) around 1987 allowed the urban constructions or real-estate developments more likely to be realized in large and regular plot inside the compact part of the city. And before that, they generally happened in a piecemeal way, mostly with small or irregular plots. Associated with that, we can generally divide rest parts of the blocks into two types. First are the parts composed by large regular plots(Figure 3, orange) and second are those with small irregular plots (Figure 3, dark grey).

The first type is easy to be understood and imagined, as they are mainly combinations of newly built residential communities with commercial area and contemporary facilities, usually with multi-story or high-rise buildings. The second group seems more illegible for us. Actually they are generally the results of the piecemeal replacement on the old settlement, the courtyard houses, whose pattern is mainly formed before the separation of usufruct right (land-use right) and linked with the history of socialist transformation of the old houses and lands. This kind of area is usually hybrid area with old houses and few small scale factories and multi-stories residences scatted among them. And Hehuatang area is just an example of this type.

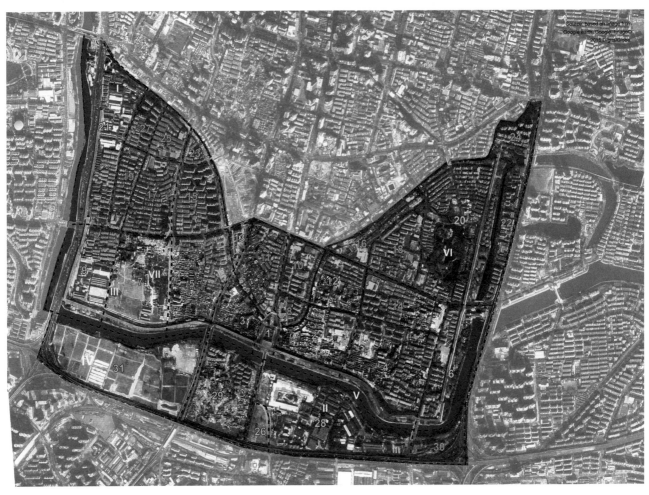

图 3 南京城南城市形态
来源：作者，基于 2015 谷歌地图图像制作

Figure 3 Urban form of South Nanjing
Source: by author, based on image from Google earth, Google, Image© 2015 DIgitalGlobe.

注释

1 2015 年荷花塘历史文化街区保护规划

2 实际上，边缘带理论首先是被路易斯·赫伯特（1936）提出来的，基于他通过柏林的城市过程观察到的一种现象：柏林的生长和物理形态，尽管由住宅的向外扩展主导着，却被一些停滞所打断，在这些时期其他种类的用地，带着其他类型的物理特征，在建成区的边缘地带被创造／建造出来。在住区扩展再次开始后，这些异类用地被嵌入城市地区之中，并倾向于保持为一个连续或不连续的区域——一个边缘带，这标志着城市历史中的一个时期，以居住区的少扩展或无扩展为特征（LOUIS，1936；WHITEHAND, GU, WHITEHAND，2011：42）。因此，我们可以知道边缘带的形成与住宅向外扩展的停滞期有关。然后这一边缘带理论被康泽恩进一步发展，他进一步解释了不同的边缘带，并将内部第一层定义为内边缘带。

3 1987 年以前，中国的土地管理制度通过行政命令划拨土地使用权，免费使用。使用者不可转让他的权利。至 1987 年，使用权与所有权逐渐分离。作为土地所有者的国家将使用权转让给特许权人一定年限，特许权人则向国家支付一定费用。这种使用权是一种财产权，可以合法转让、租赁或抵押。目前，除了通过行政分配划拨的用于军事用途、公共基础设施等公益目的土地外，其他用途的土地使用权只能通过付费获取（联合国粮食及农业组织，2004：32-33）。

Notes

1 The conservation plan of Hehuatang, 2015 http://www.njghj.gov.cn/NGWeb/Page/Detail.aspx?InfoGuid=a725d883-bf28-46d8-99bf-4d719df6f7c4

2 Actually, the fringe belt theory was firstly brought forward by Louis HERBET (1936) as he had observed some phenomenon through the urban process of Berlin that: The growth and physical form of Berlin, though dominated by the outward spread of housing, were punctuated by standstill when other types of land use, with other physical characteristics were created at the fringe of the built-up area. After renewed residential growth these land uses became embedded in the urban area and tended to remain as a distinctive continuous or discontinuous zone—a fringe belt—marking a phase in a city's history characterized by little or no extension of the residential area (LOUIS, 1936; WHITEHAND, GU, WHITEHAND, 2011: 42). From that, the relation between the standstill of residential outward spread and the fringe areas is generally explained. Then this fringe-belt concept was developed by M.R.G Conzen (Conzen, 1960) as he further explained the different fringe belts, and defined the first layer as the inner fringe belt.

3 Prior to 1987, the Chinese land management system allocated land for construction through administrative decree and free of charge. It was not possible for the user to transfer his usufruct right. As of 1987, the usufruct right has been progressively separated from the ownership right. The State, as the owner of the land, assigns the usufruct right for a certain number of years to the concessionary, who pays a fee to the State. This usufruct right, which is a kind of property right, may be legally transferred, leased or mortgaged. Presently, except for land assigned for military purposes, public infrastructure and other public welfare activities through administrative allocation, the right to use land for other purposes can only be obtained against payment of a fee (Food and Agriculture Organization of the United Nations, 2004: 32–33).

References

[1] CONZEN M R G. Alnwick, Northumberland: a study in town-plan analysis. London: Institute of British Geographers, 1960.

[2] Food and Agriculture Organization of the United Nations, F. . Multilingual thesaurus on land tenure, Chinese version, 2004. ftp://ftp.fao.org/docrep/fao/005/x2038c/x2038c00.pdf.

[3] LI G, ZHOU L. Dangdai Nanjing chengshi guihua sishi nian [Forty Years Urban Planning of Contemporary Nanjing I] (in Chinese). Modern Urban Research, 1995(1): 7–12.

[4] LI G, ZHOU L. Dangdai Nanjing chengshi guihua sishi nian [Forty Years Urban Planning of Contemporary Nanjing II] (in Chinese). Modern Urban Research, 1995(2), 50–54.

[5] H Louis. Die geographische Gliederung von Gross-Berlin[The geographical structure of Greater Berlin] // LOUIS H, PANZER W. Landerkundliche Forschung: Festschrift zur Vollendung des sechzigsten Lebensjahres Norbert Krebs. Stuttgart: Engelhorn, 1936: 164-171.

[6] WHITEHAND J W R, Gu, K, WHITEHAND S M. Fringe belts and socioeconomic change in China. Environment and Planning B: Planning and Design, 2011, 38(1), 41–60. http://envplan.com/epb/fulltext/b38/b36104.pdf.

[7] WHITEHAND J W R, Gu, K, WHITEHAND S M, et al. Urban morphology and conservation in China. Cities, 2011, 28(2): 171–185. [2010-12]. http://doi.org/10.1016/j.cities.

[8] Xie Y, COSTA F J. Urban design practice in socialist China. Third World Planning Review, 1991, 13(3): 277–296.

02

教学设计
Teaching Proposal

设计工作坊的主要目的是练习使用城市形态学和建筑类型学作为阅读和设计的工具（对于本科生），也整合涉及复杂性的一些方面，讨论类型的变化以及与社会学、经济学、可持续标准相关联的设计过程（对于硕士生）。

设计工作坊将面对南京城墙界线内部的一个地区，运用意大利传统中萨韦里奥·穆勒托里、奇安弗兰科·卡尼吉亚、阿尔多·罗西的方法学工具，尝试对中国当代城市的片段进行更新（这个教学基于都灵理工大学的教学经验，通过布置不同的任务，提案可以由本科生和硕士生共享）。

核心关注点是位于古城墙内侧的南京老城南地区的形态、环境（以及社会和经济）价值。今天的荷花塘地区被工人住宅为主的破败房屋占据，正在等待一个更新方案。这种更新也许可以不同于那些由无名高层组成的寻常街区或者所谓延续传统风格的伪复原项目。

设计工作坊的工作通过每周对设计进度和设计工作坊主题进行集体讨论组织起来。

设计过程被确定为一个渐进式的、基于步骤的过程，涉及不同的阶段，如文脉分析、前景设想、多设计选项深化、设计方案选择和贯彻完成。

The main objective of the Design Studio is practicing urban morphology and buildings typology as reading and designing tools (for bachelor students), even collecting some aspects of complexity concerning types mutations, design process in connection with society and economy, sustainability criteria (for master students).

The Design Studio will face an area in Nanjng (inside the city-walls precinct), in order to use methodological tools (in the Italian tradition of Saverio MURATORI, Gianfranco CANIGGIA, Aldo ROSSI) to regenerate parts of the Chinese city of nowadays (this proposal, based on a teaching experience at Politecnico di Torino, could be shared among bachelor and master students, giving them different deliveries).

The core of interest is the morphological, environmental (and also social and economical) assets of an area in the south part of Nanjing, located at the interior side of the ancient city-walls. The Hehuatang area (area of the Lotus Pond) has been occupied by a workers' houses slum and it is waiting today for a regeneration project, that could be different from the ordinary settlement made by tall anonymous buildings and also from the "fake" restorations in Chinese traditional style.

The work in the Design Studio will be organized by weekly collective discussions about design development and Design Studio topics.

The design process will be defined as an incremental step-based process which involved different stages, such as: diagnosis of the context, envisioning exercise and development of a number of design options, selection of a design alternative and implementation.

第 1 周：课程主题介绍（本科生 + 硕士生）
"城市阅读"：南京、中华门和城墙
（南京老城南形态分析：地图、图片、图解）

1st Week: INTRODUCTION TO THE TOPIC (B.S. + M.S.)
"Reading a City": Nanjing, Zhonghua Gate and the City Walls
(morphological analysis of southern historical Nanjing: maps, pictures, diagrams)

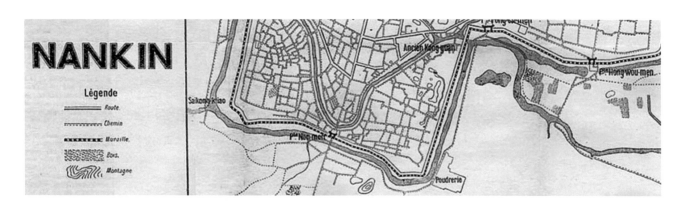

第 2 周：城市形态学（本科生 + 硕士生）
南京荷花塘地区：历史布局模式、"贫民窟"和现有更新方案
（城市尺度模型，按照"再利用"和"拆除"分类标记原有建筑）

2nd Week: URBAN MORPHOLOGY (B.S. + M.S.)
Hehuatang area in Nanjing: the historical patterns, the "slum" and the existing project
(model at urban scale, marking the original buildings to reuse and the ones to demolish)

第 3 周：原型 1.0（本科生 + 硕士生）
建筑类型：一个概念（它是什么？）和一个设计工具（如何使用？）
（中国东南部地区院落住宅及其当代诠释）

第 3 周：类型变化（硕士生）
关于庞贝（意大利古罗马城市）的练习
（一座火山城的形态和罗马"住宅"的发展）

3rd Week: PROTOTYPES 1.0 (B.S. + M.S.)
Architectural Type: a concept (what's it?) and a design tool (how may I use it?)
(the southeast China's courtyard-house and its contemporary interpretations)

3rd Week: THE VARIATIONS OF TYPES (M.S.)
An exercise about Pompei (ancient roman town in Italy)
(the morphology of a volcano's city and the development of the Roman "domus")

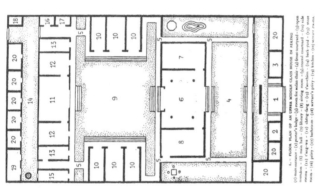

第 4 周：城市密度（本科生 + 硕士生）
相同的密度（作为城市聚居区的"数量性"特征）可以带来不同的城市设计结果（作为"质量性"产品）

第 4 周：密度与复杂性（硕士生）
不止于城市形态：在数量和质量之间的社会与经济场景
（两种不同场景对更新设计的影响）

4th Week: URBAN DENSITY (B.S. + M.S.)
The Same Density: (as quantitative feature of urban settlement) can bring to different urban design outcomes (as qualitative product).

4th Week: DENSITY & COMPLEXITY (M.S.)
Not only Urban Form: sociological and economical scenarios between quantity and quality
(the impact of two different scenarios on the regeneration design)

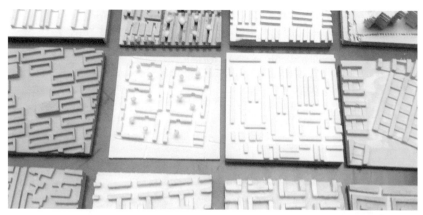

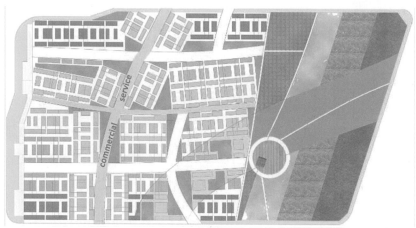

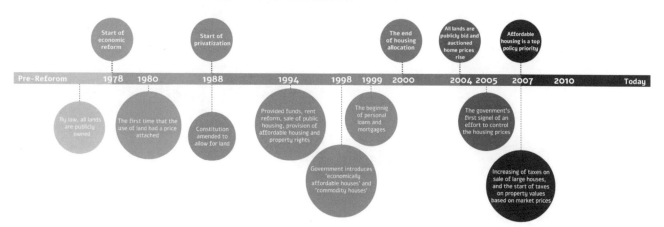

第 5 周：集群模式（本科生 + 硕士生）
应用"原型 1.0"设计聚居地
（"街区"尺度方案设计）

第 5 周：可持续性（硕士生）
传统的生态效率策略和标准化建筑
（城市尺度方案设计）

5th Week: CLUSTER PATTERNS (B.S. + M.S.)
Applying "Prototypes 1.0" in designing a settlement
(the project at "block" scale)

5th Week: SUSTAINABILITY (M.S.)
Traditional eco-efficiency strategies and standardized architecture
(the project at urban scale)

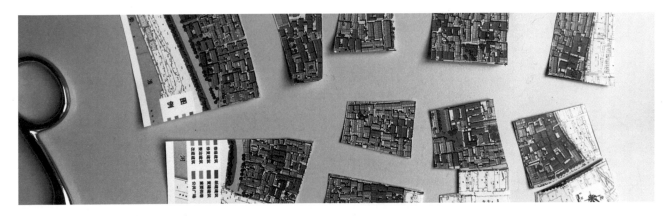

第 6 周：方案深化工作（本科生 + 硕士生）
6th Week : DETAILED SCHEME (B.S. + M.S.)

第 7 周：原型 2.0（本科生 + 硕士生）
建筑调整，新院落住宅作为矩阵构成新聚居地
（建筑尺度方案设计）
7th Week: PROTOTYPES 2.0 (B.S. + M.S.)
Tuning the architecture. The new courtyard-houses as the matrix for a new settlement
(the project at building scale)

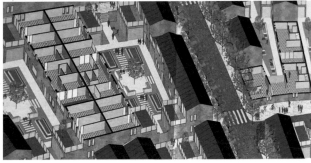

第 8 周：
最终方案成稿（本科生）
方案深化工作（硕士生）

第 9 周：
终期评图（本科生）
最后讨论与展览（硕士生）：
今日的城市形态学？

第 10 周：
最终方案讨论与深化（硕士生）

第 11 周：
终期评图与展览（硕士生）

8th Week :
Work on the Final Delivery (B.S.)
Work on the Projects (M.S.)

9th Week:
Final Presentation (B.S.)
Final Workshop&Exhibition (M.S.):
WHAT ABOUT URBAN MORPHOLOGY TODAY?

10th Week:
Reviewed and Free Work on the Final Delivery (M.S.)

11th Week:
Final Presentation & Exhibition at Distance (M.S.)

References

1. JIANG L. Morphological Research of the Historical Urban Boundary. Italy: The Inner Fringe Belt of Nanjing, PhD Dissertation, Politecnico di Torino, 2016 (tutor: Marco Trisciuoglio).
2. TRISCIUOGLIO M. Scatola di montaggio. L'architettura, gli elementi della composizione e le ragioni costruttive della forma (Toolkit. The architecture, the elements of design composition and the constructive reason of shape) Carocci, Roma (Italy), 2008, monograph translated and published in Chinese language by South East University Press, Nanjing (Popular Republic China), 2015.
3. LI X, YEO K S. Chinese Conception of Space. Beijing: China Architecture and Building Press, 2007.
4. LEDDEROSE L. Ten Thousands Things. Princeton: Module and Mass Production in Chinese Art, Princeton University Press, 2000.
5. CHEN W, GAZZOLA L. Nanjing: Comparative Study on the City Walls of Nanjing and Rome, SEU Press, 2013.
6. ROSSI A. The Architecture of the City. Cambridge Mass: MIT Press, 1982.
7. BALMER J, SWISHER M T. Diagramming the Big Idea. Routledge, London and New York: Methods for Architectural Composition, 2012.
8. PETRUCCIOLI A. After Amnesia. Bari: Learning from the Islamic Mediterranean Urban Fabric, ICAR, 2007.
9. SALAT S. Cities and Forms: On Sustainable Urbanism. Paris: Editions Hermann, 2011.

03

教案框架
Program Framework

课程开始
Course beginning

讲评 01
Review 01

讲评 02
Review 02

中期评图
Mid-term Review

本科生
Final pres
for b

| 2016.9.19 | 9.26 | 9.29 | 10.08 | 10.10 | 10.17 | 10.20 | 10.24 | 10.31 |

讲座 01
Lecture 01
南京，一个中国的城市——城市形态与城市化进程
Nanjing, a Chinese city — city form and urban process

讲座 02
Lecture 02
荷花塘，内部边界与边界再生
Hehuatang, the inner fringe and its regeneration

讲座 03
Lecture 03
类型与形态：一个观念的描述与历史
Types and Typologies: Description and history of an idea

讲座 04
Lecture 04
原型：从传统形式到新的建筑
Prototypes: From traditional type to new architecture

讲座 05
Lecture 05
模式：城市形态学，从城市分析到城市设计
Patterns: Urban morphology, from urban analysis to urban design

讲座 06
Lecture 06
建筑密度：一种理解城市的方式
Building density: A way to understand cities

讲座 07
Lecture 07
设计与制图
Designs and drawings

研究生终期评图
Final presentation for masters

| 讲评 03 | 讲评 04 | 视频交流 | | 讲评 05 |
| Review 03 | Review 04 | Skype conversations | | Review 05 |

.12 11.14   11.24   12.06   2017.1.05   1.13

讲座 08
Lecture 08

为什么要分析（物理）城市的形态？
Why analysing the morphology of the (physical) city?

―――― 本科生进度条   ―――― 研究生进度条

037

04

前期研究
Preliminary Study

练习：阅读庞贝

庞贝是古罗马的一座邻近那不勒斯的城市。公元79年，它被维苏威火山的一次突然爆发所摧毁，完全掩埋在熔岩及灰烬之下。18世纪末，考古学家开始对这个地区展开考察，使这个被一个特定的一小时悲剧所"霜冻"凝固住的城市的日常景象重现天日：人们在街道上散步，狗护卫着房屋和财产，花草树木生长在庭院与花园里。

在悲剧发生前些年的公元62年，庞贝古城发生过一次地震，震毁了几乎一半的房屋。因此公元79年的"霜冻"爆发时，这座城市正处在一个强有力的再生过程中。事实上，两次不幸的事件爆发之间的时期正是叙利亚、埃及以及地中海地区的营建风尚传入之时。这些风尚就是在乡村别墅中营造精美的花园，随后也影响到普通城市住宅的建造风气，即所谓的"居所"，由小房间环绕被称为"中庭"的庭院组成。

地震发生后，许多房主去占据房屋被破坏的场地以期扩大他们的"居所"面积，从而营建一个被柱子和花园所装饰的第二进庭院，即所谓的"柱列走廊"。庞贝在公元62年和79年间建造的这类"居所"与中国传统的两进合院建筑极其类似。虽然庞贝住宅的两进庭院是为了营造一个复杂的花园，而中国合院的第二进甚或第三进庭院是为了满足多代混居的需求：子女的家、父母的家以及祖父母的家。

观察庞贝的考古地图（一种类型地图）并且在罗马式"居所"的平面中找寻"中庭"和"柱列走廊"是一个极好的的练习，从而理解类型随时间的蜕变并领会城市形态演变的动态性。道路、入口、产权地界、庭院和房间成为这种研究最重要的空间和建筑要素。可以将一个确定的类型（即带有双庭院的"居所"）看做详细了解一个城市聚居区的关键。事实上，这是一个极富成效的认识城市的练习，并可以为后续设计奠定基础。

Exercise: Reading Pompei

Pompei is an ancient Roman city not far from Napoli. It was destroyed by the eruption of a Vulcan in 79 A.D. and suddenly and entirely covered by lava and ashes. When the first archaeologists started making research on that site at the end of XVIII century, they found a city literally "frost" in the tragedy of an hour in a determined time: the daily life of persons walking in the streets, of dogs defending houses and properties, of trees and flowers in the courtyards and in the gardens.

Few years before the tragedy, in 62 A.D., Pompei happened an earthquake that demolished half of the houses. So, the eruption of 79 A.D. "frost" a city already involved in a strong regeneration process. In fact the years between the two unlucky events were the years of a new fashion coming from Syria and Egypt and the Eastern part of the Mediterranean Basin. It was the fashion of creating elegant gardens inside the countryside villas, but that meant transfer that taste to the ordinary residential houses in the city: the so called "domus", with small rooms surrounding courtyard called "atrium".

After the earthquake, a lot of owners, decided to occupy the plots where the house have been destroyed, to enlarge their "domus" in a wider surface, creating a second courtyard, decorated with columns and adorned with gardens (the so called "peristylium"). The "domus" in Pompei, after 62 and before 79 A.D. seems an extraordinary copy of the double-courtyard house in traditional Chinese architecture, even if in Pompei the second courtyard was to create a sophisticated garden, while in China the second courtyard (as well as the third one) is to answer to the request of a "multy-generations" house: the family of the daughter, the family of the mother, the family of the grandmother.

Watching the archaeological map (a typological map) of Pompei and looking for "atrium" and "peristylium" of any Roman "domus" is an excellent exercise to understand the permutation of types in time, to realize the dynamics of urban morphology development. The roads, the entrances, the boundaries of each property, the courtyards and the rooms become the most important spatial and built elements of this kind of research. Keeping a determined type (the double-courtyard "domus") as a key to understand the detail of an urban settlement can be, in fact, a fruitful exercise to recognize urban factory and to establish fundament for the design activity.

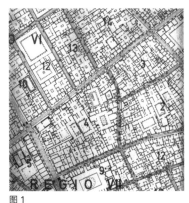
图1

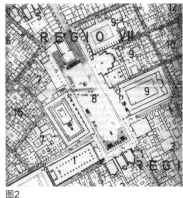
图2

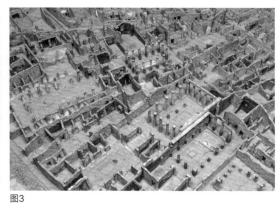
图3

图4

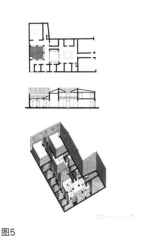
图5

图6

Figure Notes:

1. Pompei. Hof, Strasse und Platz. Stadtplan von Pompei in Masstab 1:1000 von Hans Eschebach, in Mitteilungen des Deutschen Archäologischen Insituts Römische Abteilungen, Siebzehntes Ergänzungsheft, F.H. Kerle VERLAG, Heidelberg 1970.

2. Pompei. Hof, Strasse und Platz. Stadtplan von PompeJi in Masstab 1:1000 von Hans Eschebach, in Mitteilungen des Deutschen Archäologischen Insituts Römische Abteilungen, Siebzehntes Ergänzungsheft, F.H. Kerle VERLAG, Heidelberg 1970.

3. Pompei city's model, Museo Acheologico Nazionale di Napoli

4. Pompei. General View of the Thermae, in Pompeiana: the topography, edifices and ornaments of Pompei: the result of excavations since 1819 by Sir William GELL, London 1819, Volume I.

5. House of Tragic Poet in Pompei, drawing by Claudia Sansò, in Federica Visconti, Pompeji: Città moderna. Moderne Stadt, Ernst Wasmuth Verlag, Tübingen-Berlin 2017.

6. Pompei. Par of the street of the Mercuries and the adiacent houses, in Pompeiana: the topography, edifices and ornaments of Pompeii: the result of excavations since 1819 by Sir William GELL, London 1819, Volume I.

道路和入口
Road and Entrance

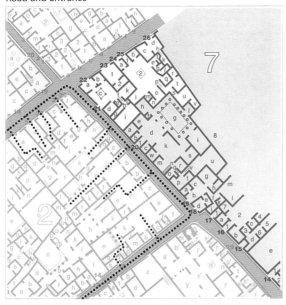

家庭户边界
Household Line

房屋产权
Ownership of Housing

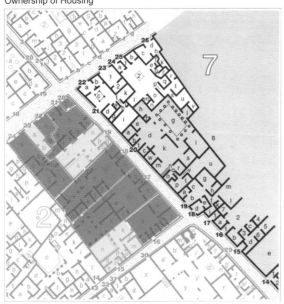

庭院和延伸
Courtyard and Extension

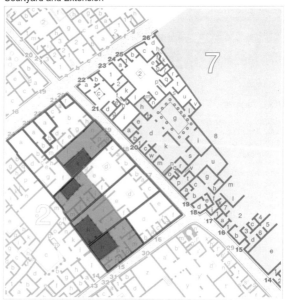

将基地分成四个相对私人的部分
Dividing the Site into Four Relative Private Parts

边界
Boundary

房间
Rooms

内部花园
Inner gradens

内部庭院
Inner yards

通道 / 走廊
Passages/Corridors

楼梯
Stairs

05

设计成果
Design Works

屋顶平面图 | Roof Plan

城市盒子

方案试图用承载着街区记忆的城市公共空间串接起荷花塘片区。基于对原街巷名称、肌理的尊重保留，用一条中心的主要道路和几条支路形成场地的主要步行系统，同时划分出几个片区。主要街巷节点处形成一系列围合的城市共享空间——城市盒子，作为对"院落"这一传统内向性共享空间在城市尺度的转译，以期让一些传统的日常生态亦可借此得到转移与延续。同时，通过对荷花塘原有建筑类型的研究，发展出适应场地的新类型并重新置入，与老建筑共存共生。

Urban Chamber

The group tried to connect the Hehuatang area with a series of urban public spaces, which carries the memory of history. Based on the original names and network of streets, they formed the major pedestrian system with a main road and several branches, dividing the area into different parts. Several streets nodes become a series of enclosed urban public spaces, named "city chambers" as urban reinterpretation of courtyard, the traditional inward-directed shared space. The expectation is that parts of the local daily life and behaviors could be transferred and kept down in such spaces. At the same time, after the study of building types in Hehuatang area, the group developed new types to get them embedded into the site, considering their compatibility and complementarity with the old buildings.

丁园白
Yuanbai DING

顾家铭
Jiaming GU

王安安
An'an WANG

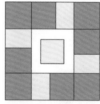

朱鹏飞
Pengfei ZHU

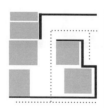

类型调研 | Typological Survey

调研现场 | Survey Scene

沿街剖面 | Sections along Street

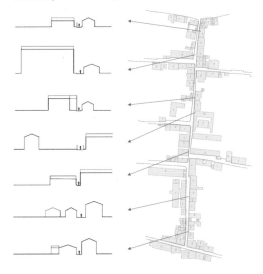

新类型 | New Type (Prototype)

类型提取 | Type Abstraction

传统庭院类型 | Traditional Courtyard Type

新类型发展 A | New Type Development A

新类型发展 B | New Type Development B

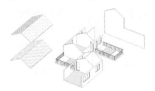

聚落布局 | Settlement Layout

组织元素 | Organization of Elements

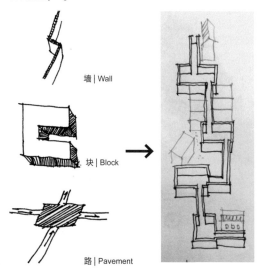

墙 | Wall

块 | Block

路 | Pavement

总体街区布置 | Layout of Block

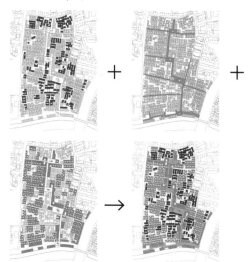

建筑密度 | Building Density

既有城市肌理 | Existing Urban Fabric

52 740 ㎡

在荷花塘截取一块区域，将历史建筑、现状建筑、主要街道和地块内小路标示清楚。根据占地面积、建筑高度等信息计算得出既有建筑密度。
Intercept a region in Hehuatang area, mark the historical buildings, the current constructions, the main streets and path in plots clearly. And then calculate the density of existing buildings. according to floor areas and building height etc.

原型模式 | Prototype Pattern

53 050 ㎡

根据自己对该地区建筑原型的分析，在保留历史建筑的前提下，用新的建筑类型合理地布置同一块基地，并计算出新的建筑密度进行比较。
According to the analysis of prototype in the region, in the premise of the historical buildings' preservation, make a reasonable layout of the same base. Calculate the new building density.

备选模式 | Alternative Pattern

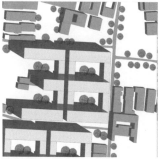

52 920 ㎡

在前一步的基础上放大原型的尺度，寻找基地布局新的可能性。在密度相差不大的情况下，集约型的建筑和留出的公共空间会带来不一样的空间效果。
On the basis of the previous step, enlarge the scale of prototype and look for new possibilities for the layout. In the case of the same density, intensive architecture and reserved public space will leave a different space effect.

049

局部放大平面 | Enlarged Plan

轴测图 | Axonometry

050

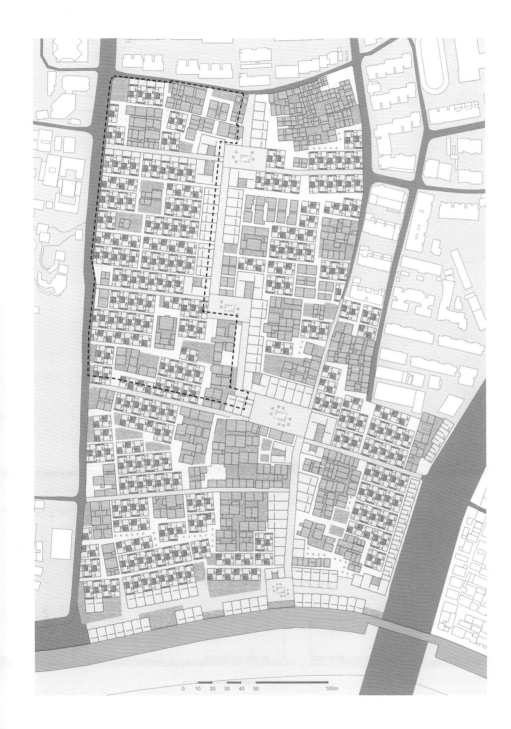

布局规则 | The Rules

1. 街区内道路均为步行道
   All roads for pedestrian paths
2. 建筑最高为两层
   Max height of 2 floors
3. 以居住功能为主，沿主街布置小商业
   Mainly residential function and small shops along the street
4. 为历史建筑、保留树木留下绿地空间
   Green space between blocks & protection of existing trees
5. 限制中心广场和主街的尺度
   Central square (maximum 40 m x 60 m) and the main road (6 m width)

总平面图 | Master Plan

双院单元一层平面 |
First Floor Plan of the Double Courtyards Unit

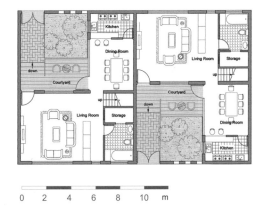

独院单元一层平面 |
First Floor Plan of the Single Courtyard Unit

街区剖面 | Cross Section

街区剖面 | Cross Section

双院单元二层平面 |
Second Floor Plan of the Double Courtyards Unit

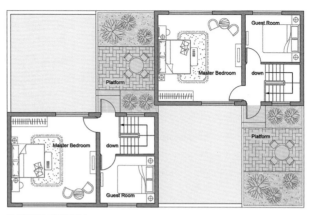

独院单元二层平面 |
Second Floor Plan of the Single Courtyard Unit

构造大样 | Construction Detail

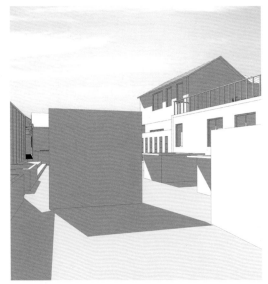

室外展场 | Outside Exhibition

茶室 | Teahouse

开放活动舞台 | Open Stage

类型调研 | Typological Survey

调研位置 | Survey Location

平面 | Plan

模型 | Model

孝顺里22号故居的入口是一个狭长的有顶走廊。走到走廊尽头就到达了一个连着两个小庭院的室外空间。这两个庭院都由几户家庭所共有。
The entrance of No.22 Houses of Xiaoshunli is a long and narrow corridor.Walk over the corridor and we will reach a outdoor space which combines two small courtyards.These two both are shared by several families.

新类型 | New Type (Prototype)

新类型 | New Type

基地特色：
老年人的日常生活：养鸟、养花、晒太阳
Features of base:
Daily life of the elderly: birds, flowers and the sun

变形参考：Deformation Reference

变形示意：
解放庭院里的加建空间，提供公共交流的中心空间。
Example of deformation:
Liberating the inserted space in the courtyard to provide public communication center space.

聚落布局 | Settlement Layout

类型变化 | Type Changes

一户变多户带来的问题：
①原先辅助空间变为主要居住空间，通风采光不满足功能需求。
②人口增多，每户用地相对变小，缺少原有合院组织的空间层次，即缺少过渡空间。
③每户的私自加建占用公共空间，使邻里交流变少。
The problem that a household becomes more complex:
① Original auxiliary space becomes the main living space, thus ventilation and lighting does not meet the functional requirements.
② The population is increasing, the land is relatively small, and there is a lack of the space level and transition space.
③ Each household's unauthorized use occupies public space, so that the neighborhood has no communication.

两户居 | Residential unit with two households

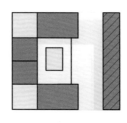

三户居 | Residential unit with three households

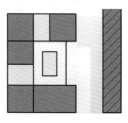

单体设计 | Single design

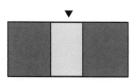

居住单元设计 | Design of residential unit

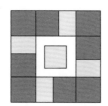

四户居 | Residential unit with four households

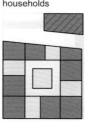

商业空间 | Commercial space

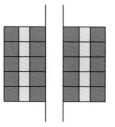

建筑密度 | Building Density

既有城市肌理 | Existing Urban Fabric

29869 ㎡

原有城市肌理呈自然生长状态。
The existing urban fabric is in state of nature growth.

原型模式 | Prototype Pattern

27877 ㎡

用新类型填充场地。
Fill the state with a new type.

备选模式 | Alternative Pattern

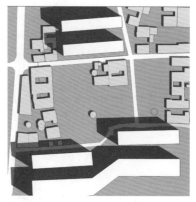

28940 ㎡

高密度状态下城市肌理。
Urban texture in high density state.

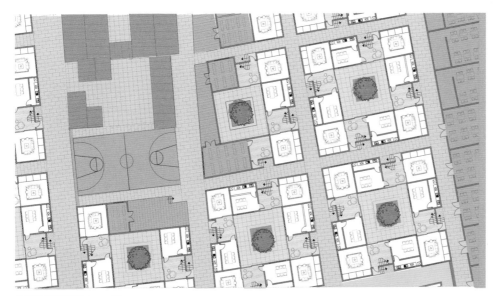

局部放大平面 | Enlarged Plan

轴测图 | Axonometry

总平面图 | Master Plan

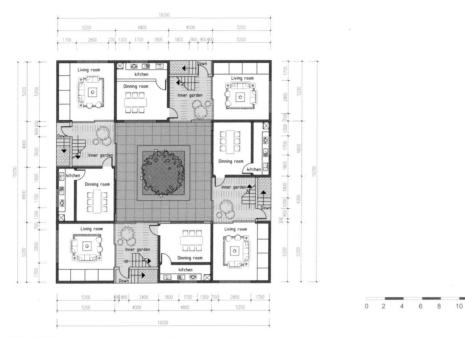

单元一层平面 | First Floor Plan of the Unit

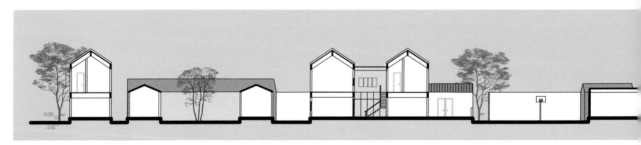

街区剖面 | Cross Section

街区立面 | Street Facade

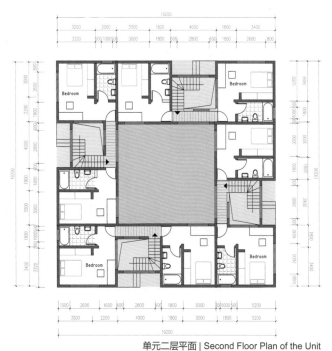

单元二层平面 | Second Floor Plan of the Unit

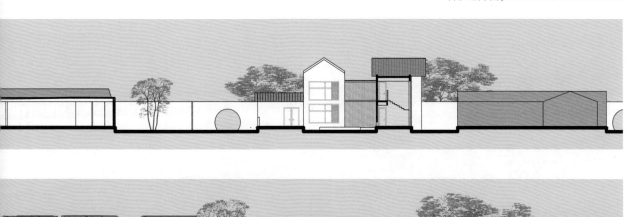

单元剖面 | Section of the Unit

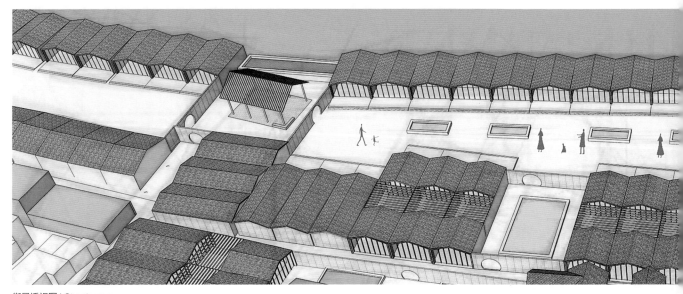

街景透视图 | Streetscape

室内场景 | Interior View

院落场景 | Courtyard View

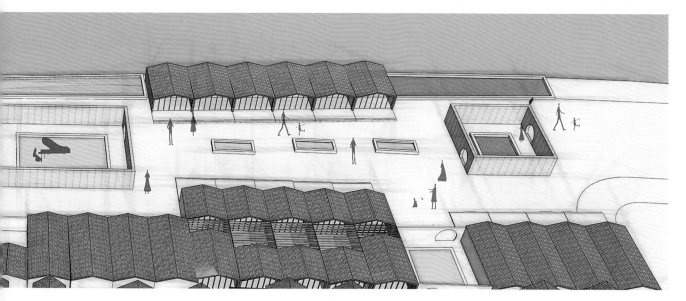

类型调研 | Typological Survey

调研位置 | Survey Location

我调研的场地是魏家烨故居,位于总场地南侧。在原有历史院落空间的基础上,由主体院落衍生出附属的小院落,具备了新的居住模式和使用功能。

My case is Jiaye WEI Old House, located in the south of the site. Based on the original historic courtyard space, the small yard is abstracted from the main yard while new, residential patterns and functions of this place are arrived.

平面 | Plan

模型 | Model

新类型 | New Type (Prototype)

新类型 | New Type

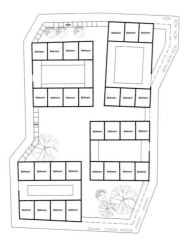

传统庭院类型 | Traditional Courtyard Type

新类型发展 | New Type Development

Re-structure—My Plan
再造—我的设计

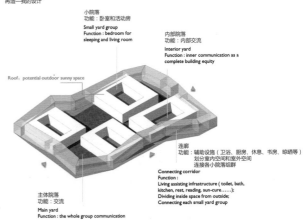

Roof: potential outdoor sunny space

小院落
功能:卧室和活动房
Small yard group
Function: bedroom for sleeping and living room

内部院落
功能:内部交流
Interior yard
Function: inner communication as a complete building equity

连廊
功能:辅助设施(卫浴、厨房、休息、书房、晾晒等)
划分室内空间和室外空间
连接各小院落组群
Connecting corridor
Function:
Living assisting infrastructure (toilet, bath, kitchen, rest, reading, sun-cure……);
Dividing inside space from outside;
Connecting each small yard group

主体院落
功能:交流
Main yard
Function: the whole group communication

新类型生成 | Process of New Type

聚落布局 | Settlement Layout

类型变化 | Type Changes

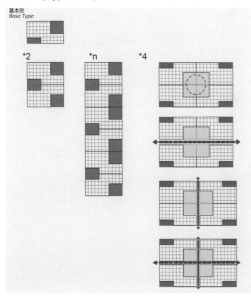

组团街巷布置 | Layout of Patterns

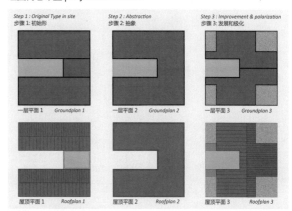

从原有场地建筑和院落的关系中重新提取基本的形态学元素，规划新的住宅院落。以两户为基本单位，形成良好的私密空间与室外景观空间。
The morphological elements are abstracted from the original site and buildings, to design the new courtyard and building type. The basic model consists of two houses. It formed good inside space and outdoor space around the house.

建筑密度 | Building Density

既有城市肌理 | Existing Urban Fabric

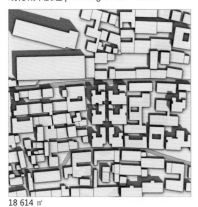

18 614 ㎡

既有城市肌理杂乱无章，改建加建情况明显。街道由建筑占据后自然形成。
The existing urban fabric of this site is disorderly and a lot of buildings are added to the empty space. The street is shaped naturally by the edge of buildings along the side.

原型模式 | Prototype Pattern

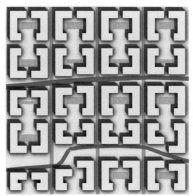

18 640 ㎡

提取新的院落模式。
More courtyard type is extracted from the site.

备选模式 | Alternative Pattern

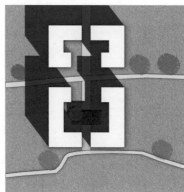

18 462 ㎡

按相同的建筑面积建造的集中的高层和周围的大片绿地。
The high building with the same building area owns the open space around it.

局部放大平面 | Enlarged Plan

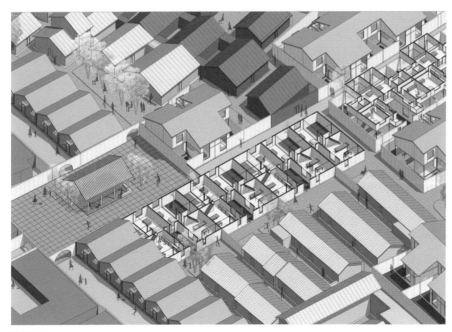

轴测图 | Axonometry

总平面图 | Master Plan

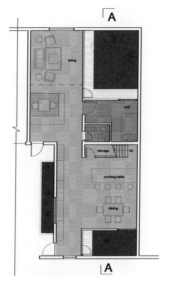

单元一层平面 |
First Floor Plan of the Unit

单元二层平面 |
Second Floor Plan of the Unit

街区剖面 | Cross Section

街区立面 | Street Facade

立面图 | Facade

剖面图 | Section

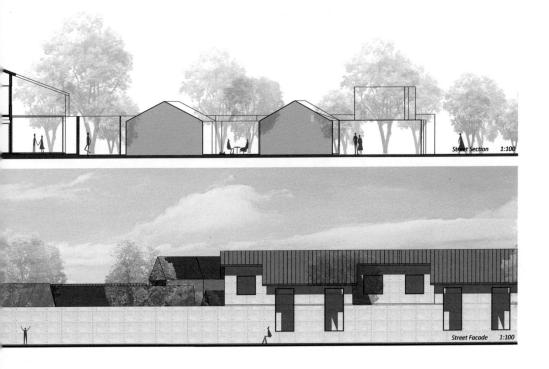

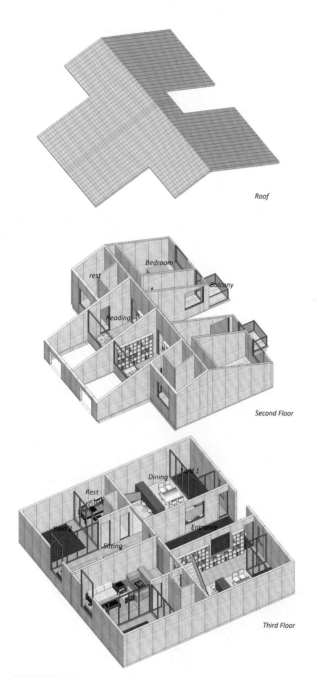

单元轴测分解 | Unit Axonometric Decomposition

室内场景 | Interior View

广场透视 | Square Perspective

071

类型调研 | Typological Survey

调研位置 | Survey Location

模型 | Model

模型 | Model

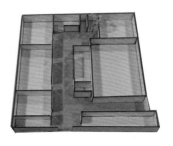

曾静毅故居有一条走廊连接院子和街道。院子被两层的历史建筑包围，中间有一个新的建筑，使得院子很狭窄。

House of Jingyi ZENG has a short corridor, which connects the street and the courtyard inside. The yard is surrounded by two-floors historical buildings and has a new-constructed one-floor building inside it, making it too narrow for activities.

新类型 | New Type Prototype

新类型 | New Type

New tpye—the father-type:
It has 3 elements (buildings, yards, small yards) and the entrance of it can be changed into a corridor or be filled in by a building. In the case, considering functions, I get 3 son-types on the left after combinations.

新类型——母类型生成三种子类型；
入口可以适应不同情况；
根据功能，我通过混合生成了三种子类型。

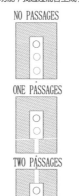

Commercial Building 商业建筑

Residential Building 居住建筑

Exhibitory Building 展览建筑

聚落布局 | Settlement Layout

类型变化 | Type Changes

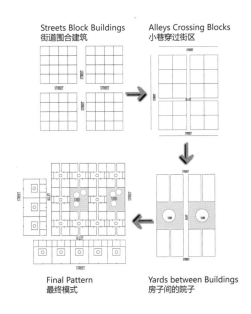

组团街巷布置 | Layout of Patterns

总体街区布置 | Layout of Block

建筑密度 | Building Density

既有城市肌理 | Existing Urban Fabric

58 500 ㎡

原型模式 | Prototype Pattern

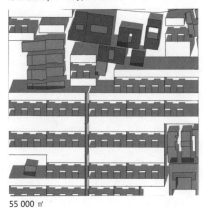

55 000 ㎡

备选模式 | Alternative Pattern

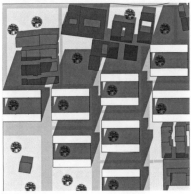

52 200 ㎡

073

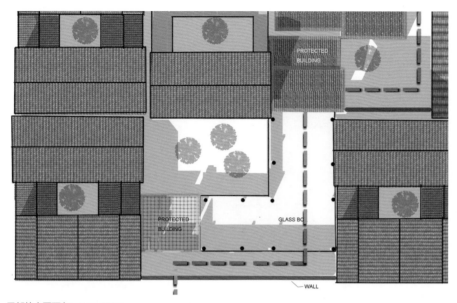

局部放大平面 | Enlarged Plan

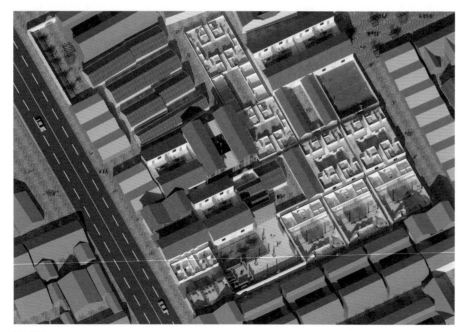

轴测图 | Axonometry

Wall、Block、Floor
墙、块、路

布局规则 | The Rules

1. 墙、块、路
   The wall, the block, the floor
2. 商业、居住、展览
   Commercial, residential, exhibitory
3. 母类型与子类型
   One mather-type, several son-types
4. 玻璃盒子用于展示和回应保护建筑
   Glass-Box, for displaying and echoing protected buidings
5. 不同透明度的材料适应不同功能
   Materials of different transparency for different functions masterplan

Master Plan | 总平面图

1 起居室 Living Room
2 卧室 Bedroom
3 厨房 Kitchen

居住一层平面 | First Floor Plan for Living   居住二层平面 | Second Floor Plan for Living

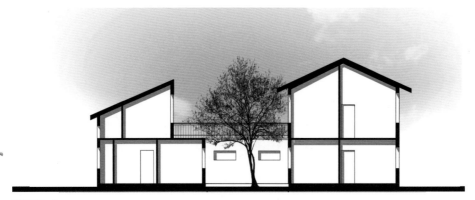

剖面 | Section

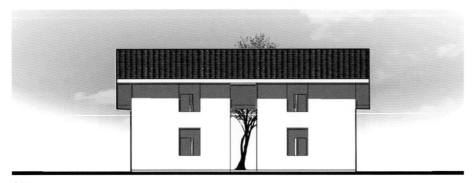

立面 | Facade

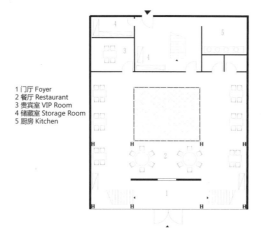

1 门厅 Foyer
2 餐厅 Restaurant
3 贵宾室 VIP Room
4 储藏室 Storage Room
5 厨房 Kitchen

商业一层平面 |
First Floor Plan for Commerce

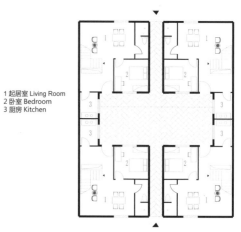

1 起居室 Living Room
2 卧室 Bedroom
3 厨房 Kitchen

商业二层平面 |
Second Floor Plan for Commerce

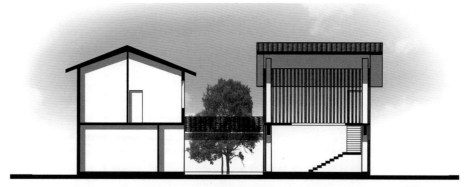

剖面 | Section

立面 | Facade

1 展示厅 Display Hall

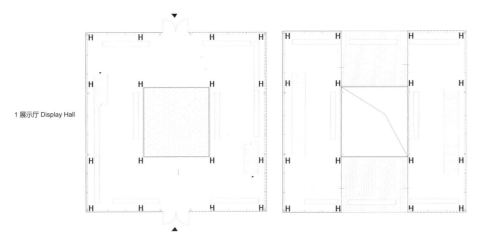

展览一层平面 | First Floor Plan for Exhibition   展览二层平面 | Second Floor Plan for Exhibition

剖面 | Section

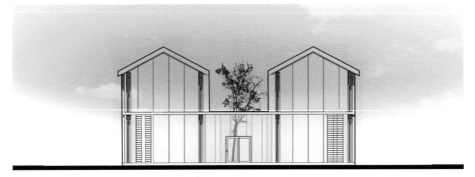

立面 | Facade

街景 | Street Views

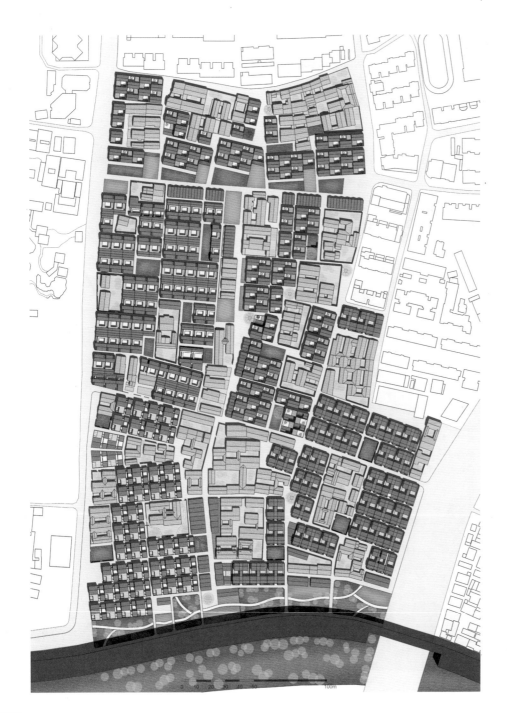

Master Plan | 总平面图 1:500

故园新生

经过与1929年航拍地图等历史信息的对比分析，我们希望提取传统的积极因素，在场地内营造公共活动的场所。整块基地围绕沿城墙和东西主要道路的两条绿带，一条南北向主轴线和一系列散布的内向型的公共花园展开。两条绿带打通愚园和中山南路的联系，南北主轴线主要承载街区商业职能，而公共花园作为具有传统意向和记忆的共享空间，为每组类型单元的原生性公共活动提供场所。

The Innovative Traditional Gardens

Reviewing the historical information such as the aerial map of 1929, the group tried to extract the effective forms and characters from tradition and create places for public activities in the field. The whole project pivots around two east-west greenbelts, one north-south main axis, and a series of inward-directed public gardens, the "innovative traditional gardens" scattered around. The two east-west greenbelts build the connection between Yuyuan and Zhongshan South road; the main north-south axis carries most commercial activities; and the "innovative traditional gardens", as shared spaces linked with traditional images and memories, provide the places of local public activities for every group of residential units.

吴舒
Shu WU

范琳琳
Linlin FAN

丁睿
Rui DING

吴和根
Hegen WU

类型调研 | Typological Survey

调研位置 | Survey Location

荷花塘 5 号民居是典型的传统三进院落民居，狭长的走道是主要的交通空间，连接两个院子。院子为房间提供良好的采光，是家庭主要的活动场所。
No. 5 Dwelling House is a typical traditional three-courtyard house. The narrow aisle is the main traffic space that connects the two yards. Courtyards are places for family gatherings, and provide favorable daylighting for rooms.

平面 | Plan

模型 | Model

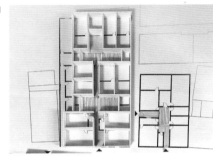

新类型 | New Type (Prototype)

新类型 | New Type

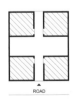 

传统庭院类型 | Traditional Courtyard Type

新类型生成 | Process of New Type

新类型发展 | New Type Development

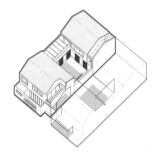

类型组合 | Type Combination

街道空间 | Street Space

聚落布局 | Settlement Layout

类型变化 | Style Changes

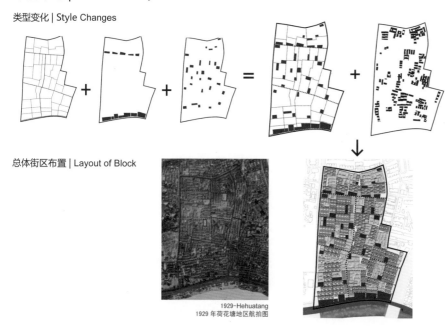

将荷花塘内部原有的道路系统与新植入的绿化景观系统整合，为整个历史街区提供全新的开放的公共活动空间。将形成的连续开放空间与历史保留建筑叠加，还原了1929年荷花塘地区的空间形态。
We integrated the original road system of this area with the newly implanted green landscape system, which provides open spaces for the whole historic district. And then we combined this new system with historical and protected buildings, getting the new district which is similar to the 1929's.

总体街区布置 | Layout of Block

1929-Hehuatang
1929年荷花塘地区航拍图

建筑密度 | Building Density

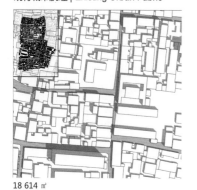

既有城市肌理 | Existing Urban Fabric

18 614 ㎡

在荷花塘历史街区中选取一块150 m×150 m的区域。
Select a 150 m × 150 m area in the historic district of Hehuatang.

原型模式 | Prototype Pattern

18 640 ㎡

保留场地中历史建筑，以合院为基本单元填满场地。
Remain the historic buildings of the site and fill the place with small courtyard houses.

备选模式 | Alternative Pattern

18 462 ㎡

保持场地容积率不变，以大体量的合院填充场地，场地内建筑密度降低，公共开放空间增加。
Then fill the place with bigger courtyard houses under the same plot ratio. The building density decreases, and this place has more public open spaces.

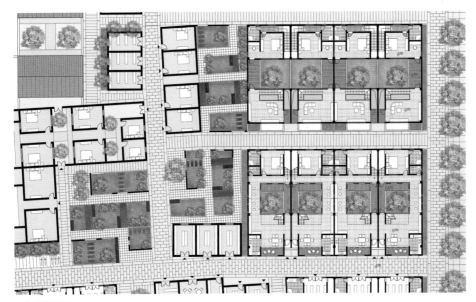

局部放大平面 | Enlarged Plan

轴测图 | Axonometry

The Innovative "Traditional Garden"
新·院

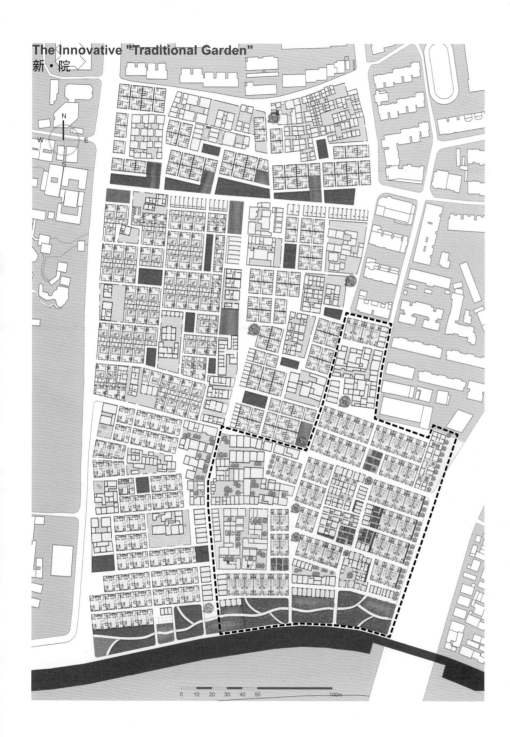

Master Plan | 总平面图 1:500

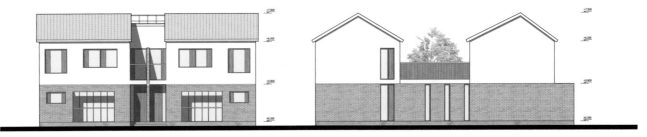

户型立面 | House Facade

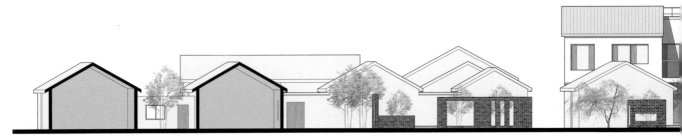

街区剖面 | Cross Section

开放空间透视图 | Open Space Perspective

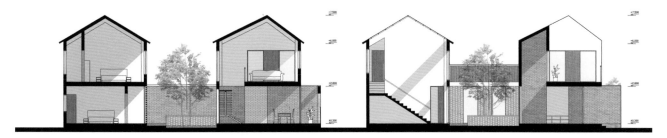

户型剖面 | House Section

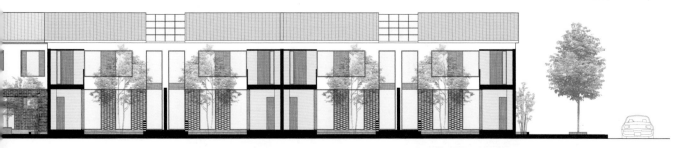

内院透视图 | Inner Courtyard Perspective

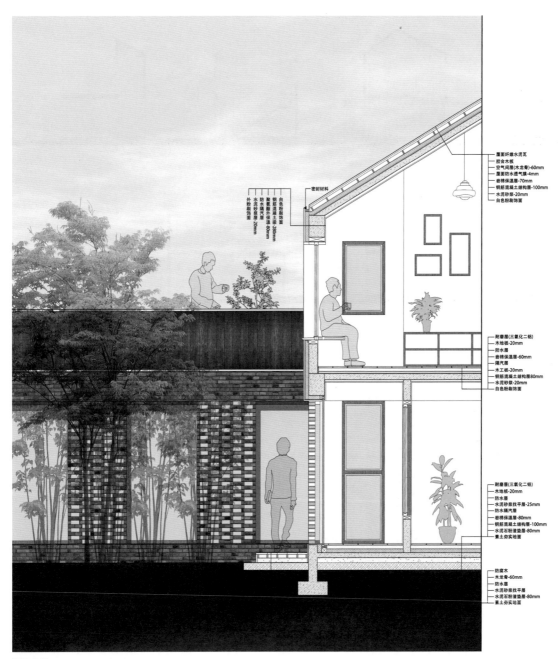

构造大样 | Construction Detail

场景透视 | Views

089

类型调研 | Typological Survey

调研位置 | Location

选取孝顺里2-6号住宅作为研究对象，位于基地的中部。这些建筑是明清遗留的保护建筑，较好地保留了当时的风貌。
We choose No.2 to No.6 residential buildings of Xiaoshun Li as the research object. These are conservative buildings of Ming or Qing dynasty and keep the style of the time.

平面 | Plan

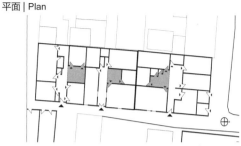

类型提取 Extract type 　　类型发展 Develop type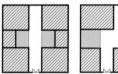

模型 | Model

新类型 | New Type (Prototype)

新类型 | New Type

荷花塘地区院落要素分析
Factors of Hehuatang courtyard

苏州院落要素分析
Factors of Suzhou courtyard

北京四合院要素分析
Factors of Beijing courtyard

新类型发展 | New Type Development

结论：新类型的四个要素
Conclusion: four elements of new type

合院作为生活中心
Yard as center of life

围墙，营造私密性
Walls, privacy

模数，便于组合
Facilitate combination

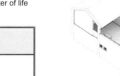

灰空间，过渡性空间
Transitional space

聚落布局 | Settlement Layout

分析1929年航拍图的场地布局可以看到成体系的公共空间，在总体设计中延续了这种理念。
从总平面的分析图中可以看出，两条公共绿带+成体系的公共空间+贯穿场地的轴线（道路联系）构成了设计的总体策略。

The systematic public space can be seen in aerial map of 1929,so in the general design we continue the old idea.
Seen from the analysis diagram of general design, overall strategy of the design constitute, two greenbelts+some systematic public space + axis through the site(the main road).

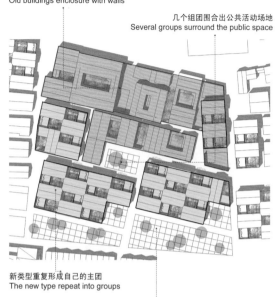

老建筑以片墙围合
Old buildings enclosure with walls

几个组团围合出公共活动场地
Several groups surround the public space

新类型重复形成自己的主团
The new type repeat into groups

公共绿带作为道路与住宅的过渡
Public greenbelt become transition of road and residence

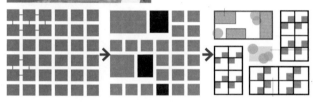

建筑密度 | Building Density

既有城市肌理 | Existing Urban Fabric

18 614 ㎡

在场地中框出30 m×30 m的范围，灰色部分为保留的历史建筑，既有城市肌理十分密集。
Choose a 30 m×30 m square on the site, the grew part are reserved historical buildings, the existing urban fabric shows a high density.

原型模式 | Prototype Pattern

18 640 ㎡

将新类型的住宅放入这一区域中，保持相同的建筑密度，场地中留出几处公共活动空间。
Replace the existing fabric by the new types and keep the same density, some public squares appear in the site.

备选模式 | Alternative Pattern

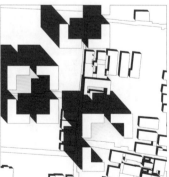

18 462 ㎡

将高层点式住宅放入这一区域中，保持相同的建筑密度，场地上空出大片绿地，但图底的感觉消失。
Replace the existing fabric by some point-type high-rise residential buildings and keep the same density, much space is released but the sense of figure ground relation disappear.

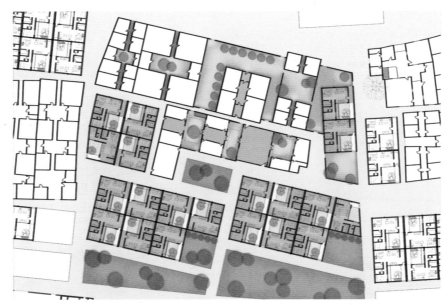

局部放大平面 | Enlarged Plan

轴测图 | Axonometry

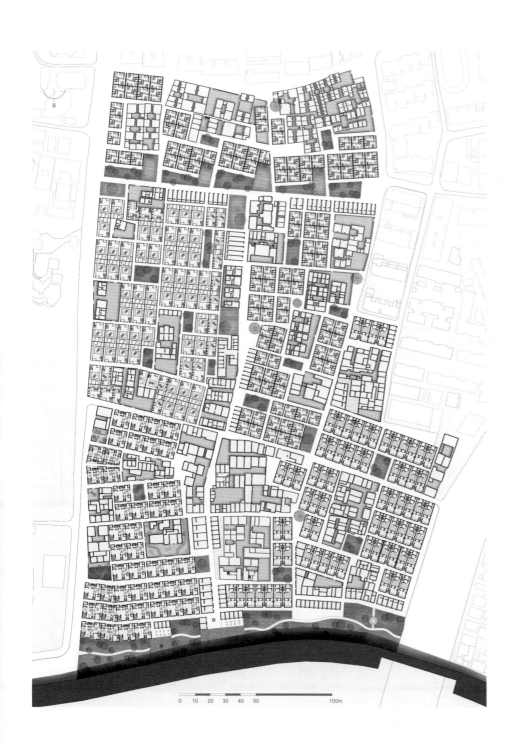

Master Plan | 总平面图 1:500

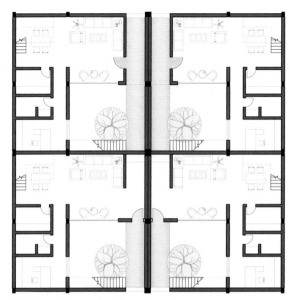

单元一层平面 | First Floor Plan of the Unit

街区剖面 | Cross Section

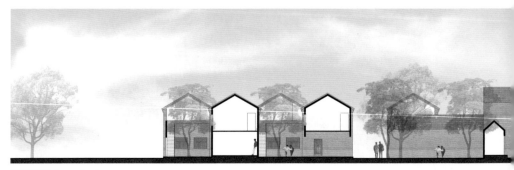

街区剖面 | Cross Section

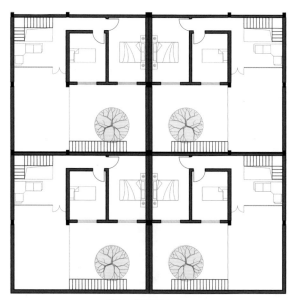

单元二层平面 | Second Floor Plan of the Unit

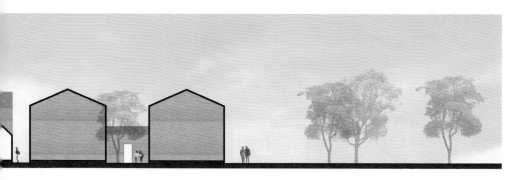

构造大样 | Construction Detail

巷弄场景 | Lane View

开放空间场景 | Open Space Views

类型调研 | Typological Survey

调研位置 | Location

平面 | Plan

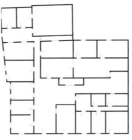
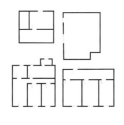

模型 | Model

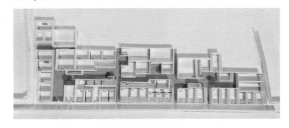

街道是我们调研的主要场所，我们发现街道大都狭窄，但是具有丰富的活动空间。院落贴着街道，使交流更加密切，活动连成一片。
The street is the main place for our research, we found that the streets are narrow but with a lot of activity. The courtyard is attached to the street, making the communication more closely .

新类型 | New Type (Prototype)

新类型 | New Type

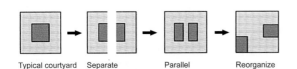

Typical courtyard — Separate — Parallel — Reorganize

传统庭院类型 | Traditional Courtyard Type

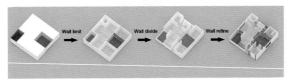

Wall limit — Wall divide — Wall refine

新类型生成 | Process of New Type

新类型发展 | New Type Development

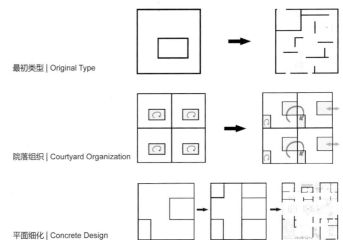

最初类型 | Original Type

院落组织 | Courtyard Organization

平面细化 | Concrete Design

聚落布局 | Settlement Layout

外墙分隔
Exterior Wall

限定空间
Limit the Space

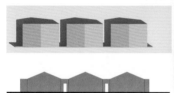 

围合空间 | Enclosed Space

结合使用
Combination

内墙布置 | Interior Wall

传统住宅中，墙用于形成一个个小房间，兼承重作用。
In traditional residential area, the wall is used for the formation of a chamber and for load-bearing.

荷花塘 | Hehuatang

巴塞罗那德国馆中，墙体被解放出来，划分空间形式更加自由，形成流动空间。
In Barcelona Pavilion, the wall was liberated, the space is divided more freely, to form flowing space.

巴塞罗那德国馆 | Barcelona Pavilion

建筑密度 | Building Density

既有城市肌理 | Existing Urban Fabric

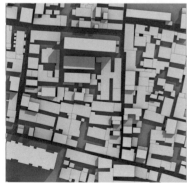

18 614 ㎡

最初的荷花塘特色肌理，是低层、高密度和自发形成的。
The original characters of Hehuatang area is low height, high density and spontaneous formation.

原型模式 | Prototype Pattern

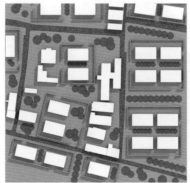

18 640 ㎡

用墙来组织院落关系，在保护既有建筑的原则上增加功能。
Use the wall to organize the courtyard and add the function of the existing buildings.

备选模式 | Alternative Pattern

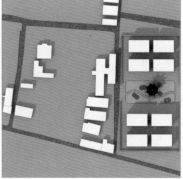

18 462 ㎡

增加楼层高度，同时使用外部围墙围合空间，形成景观与建筑相结合的情况。
Increase the height of the floor and using external walls to enclose space, which forms a combination of landscape and architecture.

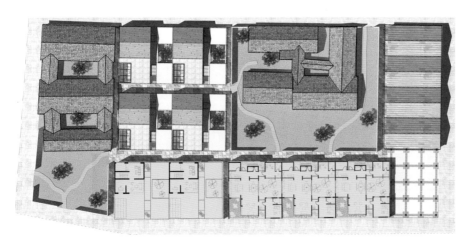

局部放大平面 | Enlarged Plan

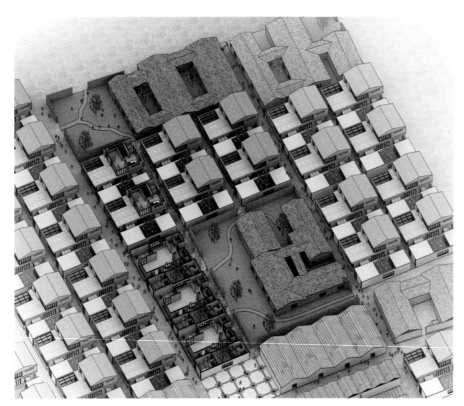

轴测图 | Axonometry

布局规则 | The Rules

1. 历史建筑 Historical buildings:
   - 较好地保存和修复
   - 用于旅游和商业

   Be well preserved and restored
   Used for tourism or commerce

2. 街区 Streets
   - 仅作步行
   - 主路：不小于 4 m
   - 住区支路：不大于 2.5 m

   Only pedestrian
   Main road : min 4 m
   Residential Lane: max 2.5 m

3. 住区 Residential area
   - 最高 2 层
   - 带院子

   Max 2 floors
   With couryards

Two green belts linked with garden and city wall

+

Commercial belts connect green belts

+

Historical buildings added with wall

↓

Road nets with historical bulildings

Master Plan | 总平面图 1:500

1	主入口 Main entrance	2	餐厅 Dining room
3	厨房 Kitchen	4	起居室 Living room
5	主卧 Main bedroom	6	卫生间 Bathroom
7	客房 Guest room	8	卫生间 Bathroom
9	储藏 Storage	10	内部庭院 Inner garden

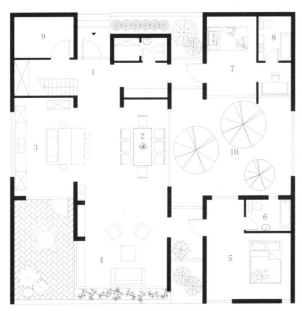

单元一层平面 | First Floor Plan of the Unit

街区剖面 | Cross Section

街区立面 | Street Facade

1 书房 Study room
2 景观房 Viewing room
3 休息室 Rest room
4 观景平台 Viewing platform

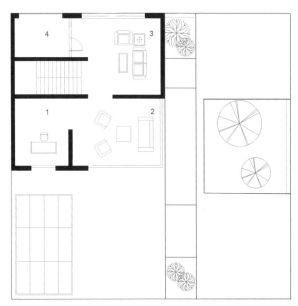

单元二层平面 | Second Floor Plan of the Unit

类型调研 | Typological Survey

调研位置 | Survey Location

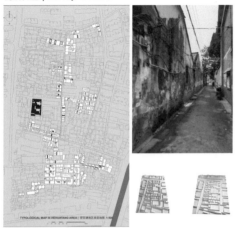

调研场地位于荷花塘地块的西部，通过对场地中所选的类型分析，去除加建，得出了所选地块建筑基本原型。
Survey site is located in the west part of Hehuatang area, according to analyse the chosen type and delete the extensions, we get the basic type of abstracted area.

平面 | Plan

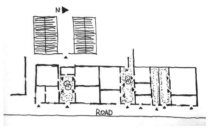

模型 | Model

新类型 | New Type (Prototype)

新类型 | New Type

新庭院类型 | New Courtyard Type

类型平面 | Plan of New Type

新类型发展 | New Type Development

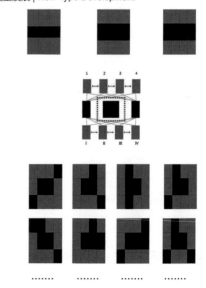

104

聚落布局 | Settlement Layout

组团街巷布置 | Layout of Patterns

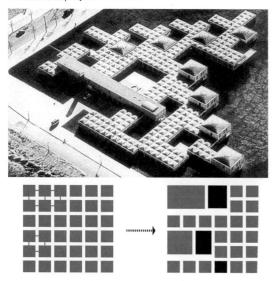

总体街区布置 | Layout of Block

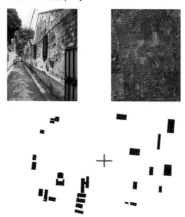

将荷花塘街区的老房子秩序和设计重新植入的"新"院秩序叠合，形成新的公共空间布局。
Combine the old house in Hehuatang with the new courtyards to form different space arrangement.

建筑密度 | Building Density

既有城市肌理 | Existing Urban Fabric

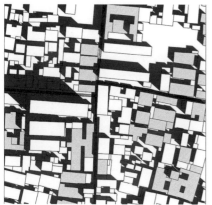

18 638 ㎡

现存的城市肌理较为混乱，光照条件较差，公共空间可达性较差。
Existing urban fabric is pretty complex, without little light, low available public spaces.

原型模式 | Prototype Pattern

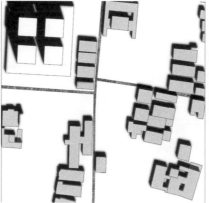

18 800 ㎡

高层可为城市空间带来较多的公共空间。
High rising building bring much more public spaces to urban space.

备选模式 | Alternative Pattern

18 500 ㎡

高密低层弥补城市肌理，也提供了怡人的围合空间。
Low rising buildings make up the urban fabric, and provide the comfortable enclosure spaces.

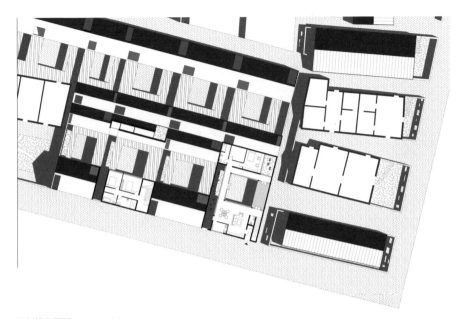

局部放大平面 | Enlarged Plan

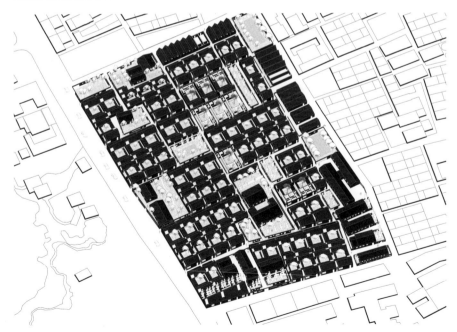

轴测图 | Axonometry

Trace 痕迹

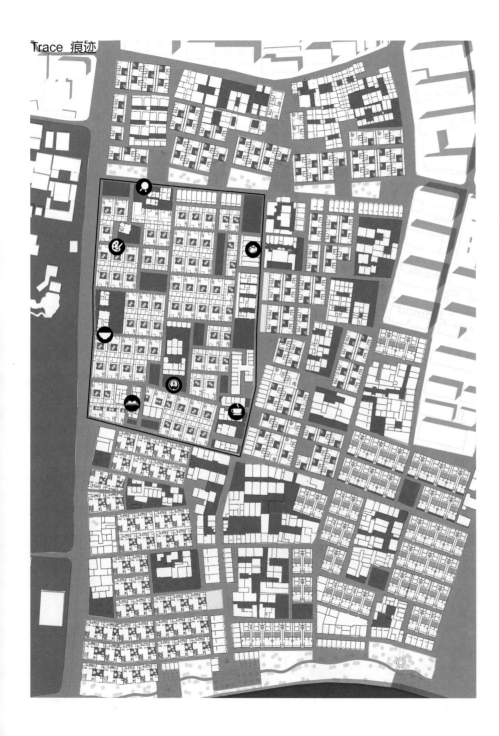

布局规则 | The Rules

1. 道路宽 2.5 ~ 3 m
   2.5~3 m road in the area
2. 带室外庭院
   Outside chamber
3. 由围墙限定
   Limited with wall
4. 连接新老景观带
   Connection: the old scenic band and the new one

Master Plan | 总平面图 1:500

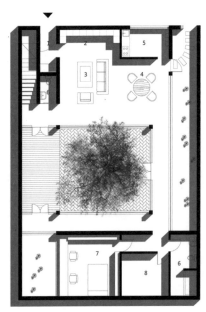

单元一层平面 | First Floor Plan of the Unit

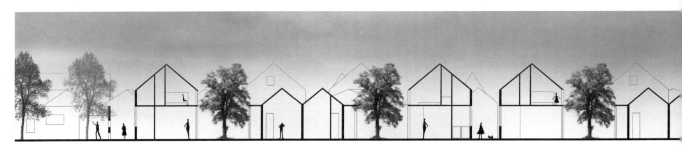

街区剖面 | Cross Section

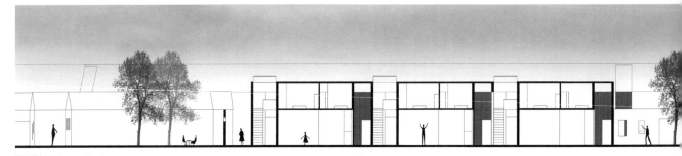

街区剖面 | Cross Section

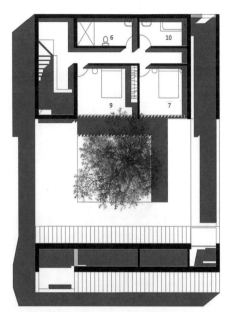

单元二层平面 | Second Floor Plan of the Unit

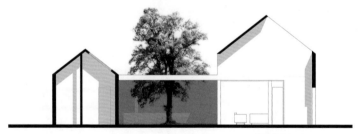

街区剖面 | Cross Section

街区剖面 | Cross Section

Roof Plan | 屋顶平面图

水井——茶室

场地中留存有多处井巷空间，可窥南京老城南传统生活方式一隅。本案以井台空间为切入点，保留水井这一传统市井生活中特有的标志性空间，并以之为契机营造场地内匮乏的公共活动空间，植入茶室等使用功能以形成活动空间节点，从而记存南京老城南的历史和传统生活的记忆碎片。本案挖掘传统生活中具有重要公共意向的水井空间的价值，试图打开尘封记忆，将之作为重塑该片区公共空间的激发点，并尝试用类型学的方法理清该片区公共空间的结构脉络，分层次讨论不同尺度下公共空间、建筑形态及人的活动之间的相互影响，探讨针对老城区渐进式更新改造的某种可能性。

施剑波
Jianbo SHI

吴则鸣
Zeming WU

Wells and Tea Houses

It is easy to find that there are still many traditional wells and alleys in Hehuatang area. They reveal people a corner of the traditional lifestyle in the Old South City of Nanjing. The well is the keyword of Group C. Maintaining and restoring all the wells, even expanding their symbolic value into urban spaces and tea houses is a way to keep some pieces of the memories of the historical and traditional life in the Old South City of Nanjing. This design work is intended to protect the remaining old wells and the memories it refers to. In this case, urban morphology and architectural typology became efficient instruments to make students clarify the existing system and the hierarchy of the public spaces in the area, facing the design question about the relationship and interplay among the public spaces, building types and human behaviors in different levels. Furthermore, it provided the student a chance to explore another potential approach for a gradual regeneration of the old town.

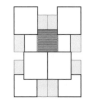 

宋梦梅
Mengmei SONG

类型调研 | Typological Survey

场地印象 | Site Impression

重构 | Reconstruction

日常 | Daily life

井边生活 | The Well, the Life

荷花塘水井现状 | Existing conditions of wells in Hehuatang area

井是公共空间的象征 | Well as the symbol of the public space

新类型 | New Type (Prototype)

调研类型 | Survey Type

建筑现状 | Existing Buildings

简化 | Simplify

 或 Or

直线形路径
Straight path

折线形路径
Zigzag path

新类型发展可能性 | Possibilities of New Type

新类型 | New Type

A

直线形路径 | Organized by the straight path
各组团共享半公共院子 | Each group share a semi-public courtyard
各房间朝向半私密院子 | Each room face towards a semi-private courtyard

B

折线形路径 | Organized by the zigzag path
各房间朝向半私密院子 | Each room face towards a semi-private courtyard

"庭院—路径"关系
Yard-Path Relationship

a. 两侧，对位
a. 2 sides, 1 place

b. 两侧，错位
b. 2 sides, 2 places

c. 一侧
c. 1 side

a. 漫步式路径
a. no main path

b. 主次分明的路径
b. one main path

聚落布局 | Settlement Layout

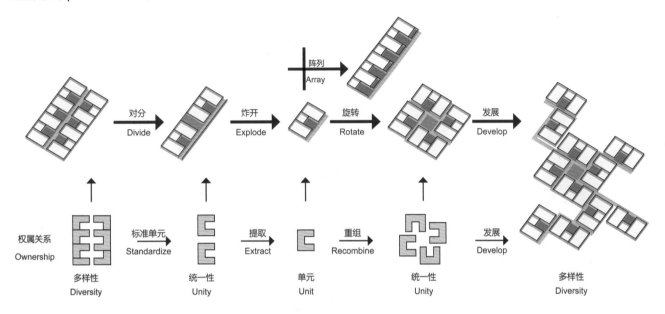

建筑密度 | Building Density

既有城市肌理 | Existing Urban Fabric

原型模式 | Prototype Pattern

备选模式 | Alternative Pattern

55 480 m²

56 630 m²

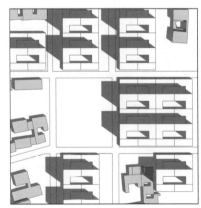

55 650 m²

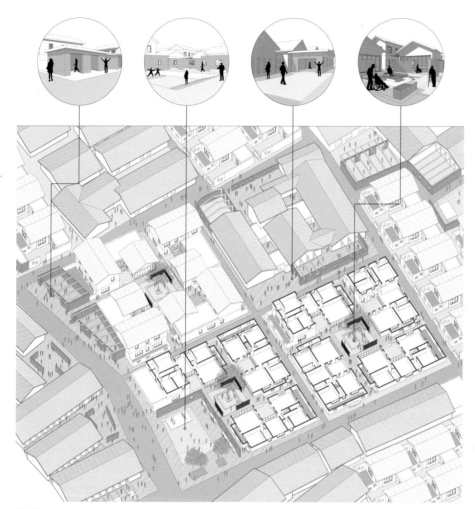

轴测图 | Axonometry

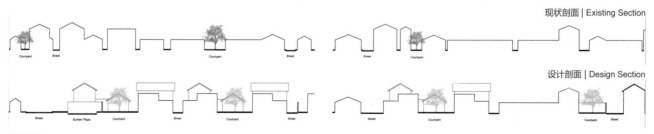

现状剖面 | Existing Section

设计剖面 | Design Section

城市剖面 | Cross Section

水井—生活
The Well, the Life

布局规则 | The Rules

1. 保留现状古井，将其发展为下沉的开放广场。
   Keep existing wells and develop them to be public sunken plazas.
2. 保留古旧建筑，将其改造以适应现代性需求。
   Keep old buildings and reconstruct them to satisfy modern needs.
3. 住区内新建水井，留存传统生活方式的记忆。
   Rebuild new wells in each new quarter to keep the memory of traditional lifestyle.
4. 通过墙体限定公共空间，与居住区相互区分。
   Use walls to form public space and distinguish from the residence.

Master Plan | 总平面图

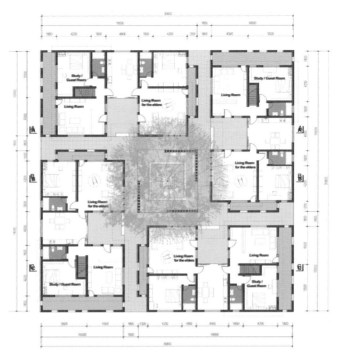

单元一层平面 | First Floor Plan of the Unit

街区剖面 | Cross Section

街区立面 | Street Facade

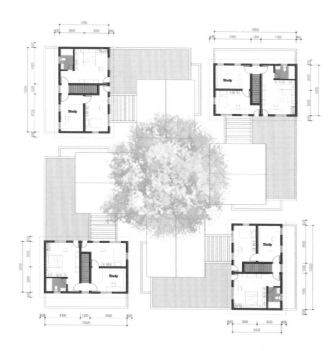

单元二层平面 | Second Floor Plan of the Unit

构造大样 | Construction Detail

井院场景 | Well-Courtyard View

类型调研 | Typological Survey

水井位置 | Location of Wells

场地中分布有多处水井，它是传统井巷生活的见证，部分井今天仍在使用。
There are some wells spread in the site, they are the witness of the Chinese traditional well-alley life, some of these wells are still been used now.

广场分布 | Layout of Squares

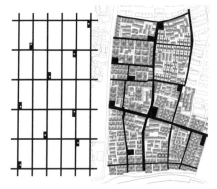

根据水井的位置分布，我们布置了几处小广场，让各个片区都拥有足够的公共空间。
According to the location of the wells, we planed some squares to let each block has their own public space.

原型调研 | Prototype Surveys

我们考察了几处居民房并制作其类型学模型，然后根据调研得出的需求进行概念化变形设计。
We researched some residential building and made some typological model, then we did conceptual design according to the need we found in the area.

新类型 | New Type (Prototype)

类型分析 | Typological Analysis

根据门窗、楼梯位置等对调研的居民房进行了类型学抽象。
We abstracted the residential building in a typological method according to the position of doors, windows and stairs.

新类型生成 | Generative Process of New Type

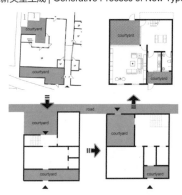

对原有类型进行一步步变形，产生了适合现代生活并保留了原有生活尺度的新类型。
Transform the prototype step by step and generate a new type which can both meet the modern lifestyle and keep the original living scale.

新类型模型 | Model of New Type

新类型拥有入户的前院，狭小的后院，起居室外的活动平台，院中的树象征家庭的精神核心。
The new type has a front courtyard for entrance, a narrow backyard, a platform outside the living room and a tree symbolize the spiritual center of the family.

聚落布局 | Settlement Layout

案例分析 | Case Study

该案例中，先由公共庭院到达各自的私人庭院，再进入各自的家中，层级分明。
This case is hierarchical in terms of the forms of the courtyard. First you come into a public courtyard, then go through private courtyards and enter home respectively.

布置原则 | Rules of Pattern Settlement

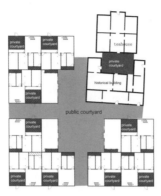

三个新建小住宅和一个商铺拼成一个住区单元，三个住区单元和一幢保留的优秀历史建筑组成一个生活单元，并围合出生活小广场。
3 residential courtyard houses and a shop combine into a block, 3 new blocks and a historical block combines into a unit and forms a public courtyard.

组团模型 | Model of a Unit

组团围合出的小广场，三面为商铺，一面为历史建筑，不让广场活动影响入户行为。
The interface of the small square enclosed by the blocks are shops on three faces and a historical building on one face and do not let the square activity intervene the entrance behavior.

建筑密度 | Building Density

既有城市肌理 | Existing Urban Fabric

在场地中框出 30 m×30 m 的范围，灰色部分为保留的历史建筑，既有城市肌理十分密集。
Choose a 30 m × 30 m square on the site, the grew part are reserved historical buildings, the existing urban fabric shows a high density.

原型模式 | Prototype Pattern

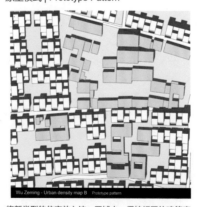

将新类型的住宅放入这一区域中，保持相同的建筑密度，场地中留出了几处公共活动空间。
Replace the existing fabric by the new types and keep the same density, some public squares appear in the site.

备选模式 | Alternative Pattern

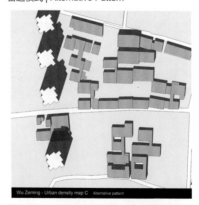

将高层点式住宅放入这一区域中，保持相同的建筑密度，场地上空出大片绿地，但图底的感觉消失。
Replace the existing fabric by some point-type high-rise residential buildings and keep the same density, much space is released but the sense of figure ground relation disappear.

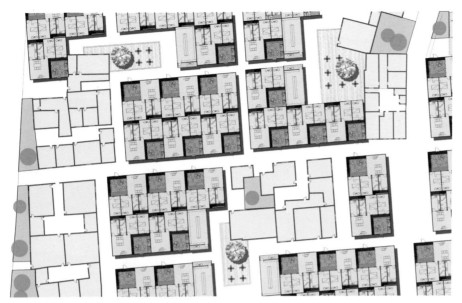

局部放大平面 | Enlarged Plan

轴测图 | Axonometry

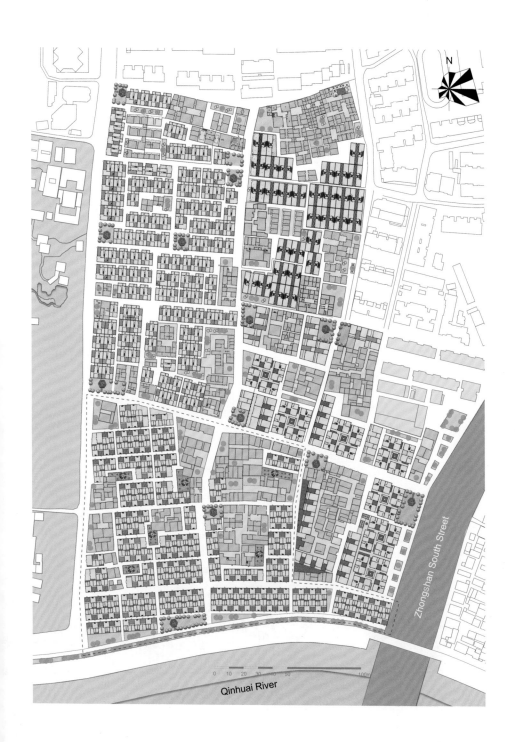

总平面图 | Master Plan

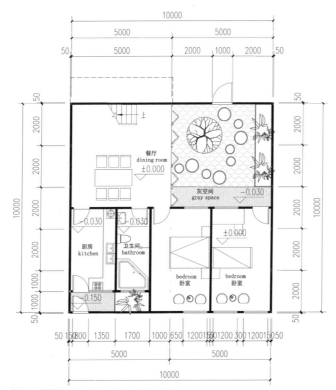

单元一层平面 | First Floor Plan of the Unit

街区剖面 | Cross Section

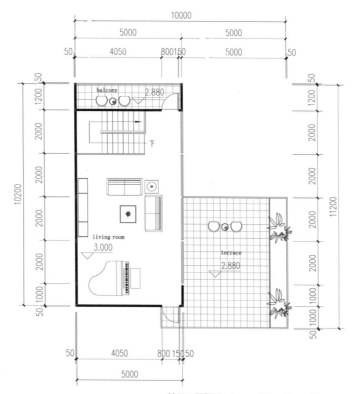

单元二层平面 | Second Floor Plan of the Unit

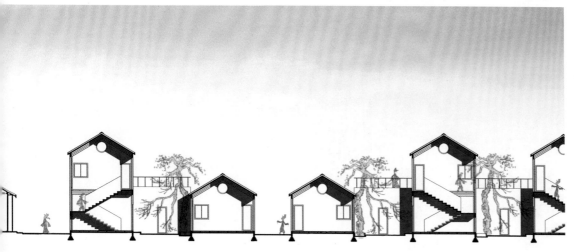

场景透视 | Views

构造大样 | Construction Detail

类型调研 | Typological Survey

调研位置 | Survey Location

我们记录了现存作为城市符号的井的位置，并发现现存的井基本上是被住宅包围用来供水
We record the location of the existing wells as a kind of urban symbols and find that the well is basically surrounded by the residences to share the water.
启示：我们可以将井用一个半公共盒子包围，使人在合适的尺度中回忆曾经的美好往事。
Inspiration: We can surround the well with a semi-open box and make people recover the moment before in a suitable-scale space.

平面 | Plan

模型 | Model

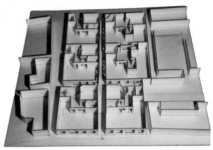

新类型 | New Type (Prototype)

新类型 | New Type

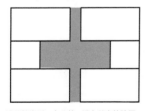

选用的类型：太公共，没有私密的院子
Type of selection: too public, no private courtyard

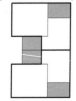

置入前院和后院来增加私密性，同时保持入口的小部分公共空间
Add front and back courtyard

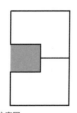

切分房屋
Split the building for two families

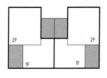

旋转90°获得南向采光
Rotate 90° to face to the south

新类型发展 | New Type Development

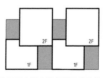

两个院子隔出的实体空间由北边的二层盒子和南边的一层盒子构成（考虑采光）
South 1F, North 2F(consider the sun)

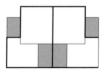

不同组合方式（对称或平行）共享后院或者共享平台
Symmetry or parallel
Shared courtyard backwards or shared platform

商住两用
1F can change into commerce when meeting the square

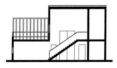

院子间是公共和核心部分，卧室朝着院子
Between the courtyards is the public and core space for the family, bedrooms face to the courtyards

聚落布局 | Settlement Layout

类型变化 | Type Changes

与现存历史建筑的关系
Relationships with the existing buildings

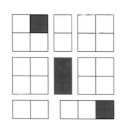

1. 住宅——成为组团的一部分
2. 公共空间——保护建筑在街区中部
1 Residence – a new part of the combination
2 Public space – in the middle of the block

组团街巷布置 | Layout of Patterns

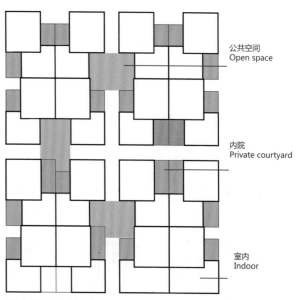

公共空间
Open space

内院
Private courtyard

室内
Indoor

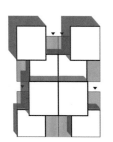

组团
Group

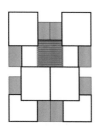

平台分公共和使用部分
Platform can be semi-open in the middle and private for the other part

单体具有可变性：院子可变成室内或者打断墙变成公共空间
Cell can change: courtyard can change into indoor or shaded open area

建筑密度 | Building Density

既有城市肌理 | Existing Urban Fabric

原型模式 | Prototype Pattern

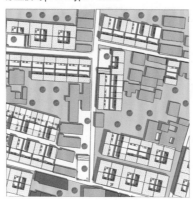

备选模式 | Alternative Pattern

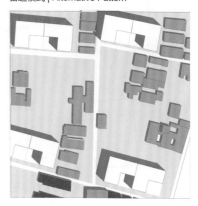

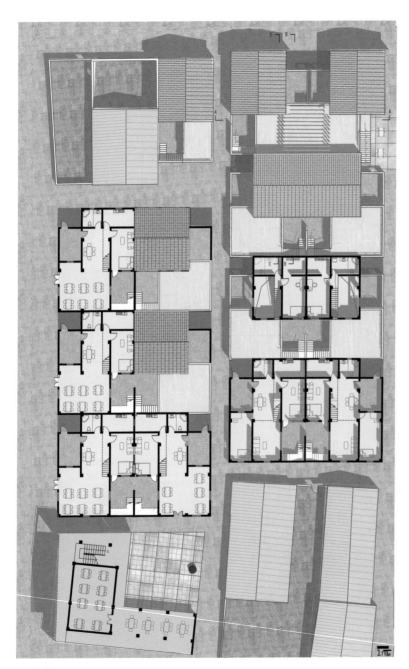

局部放大平面 | Enlarged Plan

布局规则 | The Rules

1. 茶室 Teahouse
   基于现有和新的井的位置布置茶室场地上平均分布创造新的街区网络 3 层高来欣赏整个场地
   Location based on the existing and new wells Spread evenly through the site Create new block network Three-story-high to view the whole site
2. 建筑 Buildings
   与旧建筑类似的新的形式
   历史建筑保持原来的形式
   New types stay similar to the old type Historical buildings keep their original type
3. 街道 Streets
   主要行人街道 6m 宽
   次要行人街道 4m 宽
   Six-meter-wide main pedestrian street Four-meter-wide secondary pedestrian street
4. 组团 Groups
   根据广场布置
   Arrangement according to the square
5. 现存建筑 The existing buildings
   围墙复原曾经的院子
   The wall surround them to recover their courtyard before

总平面图 | Master Plan

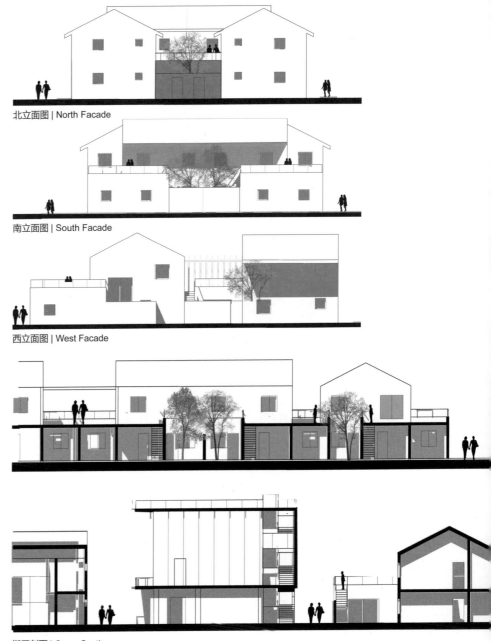

北立面图 | North Facade

南立面图 | South Facade

西立面图 | West Facade

街区剖面 | Cross Section

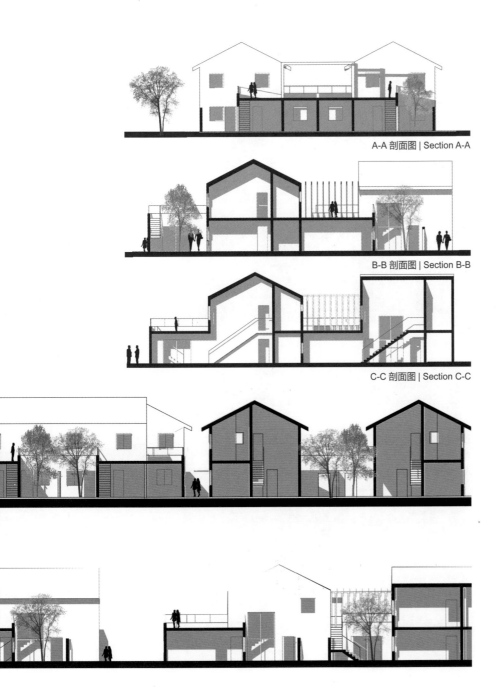

A-A 剖面图 | Section A-A

B-B 剖面图 | Section B-B

C-C 剖面图 | Section C-C

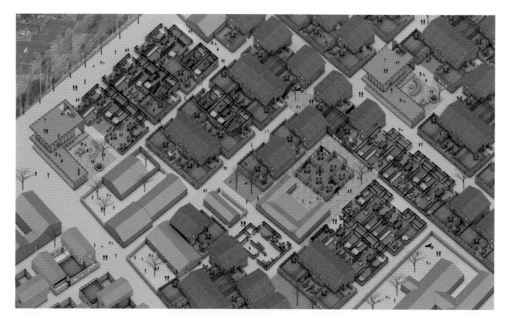

轴测图 | Axonometry

场景透视 | Views

场景透视 | Views

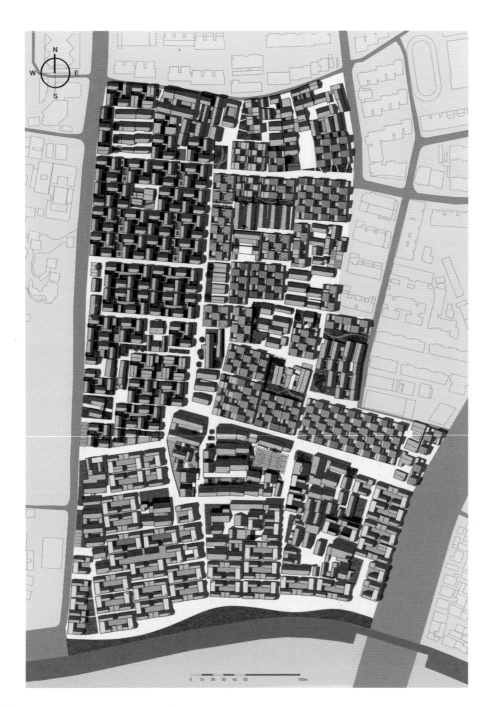

Roof Plan| 屋顶平面图 1:500

辐射

我们组的设计概念为"辐射"。通过对场地历史建筑的总平面布局的梳理与探讨，在荷花塘区域内一组保护建筑潜在的文化空间价值被发掘出来。本案将这一核心地区设计为整个荷花塘居民区的社区活动中心，空间布局上，道路的走向和尺度等级与这组中心历史建筑相呼应。方案最终生成一个放射性的街道和空间系统，以接纳并引导新的建筑布局。

刘巧
Qiao LIU

韩珂
Ke HAN

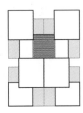 

黄一凡
Yifan HUANG

Beaming

The keyword of this project is beaming. Studying and redrawing the map of the historical settlement gives a outcome that the potential role of a group of protected buildings of Hehuatang area is discovered. Project aims to make this core become the district community center, so that the orientation and hierarchy of the streets are traced in accordance with the position of that central group of historical buildings. Finally a radial system of streets and spaces has been generated by the project to allow and suggest the design of new buildings' patterns.

类型调研 | Typological Survey

调研位置 | Survey Location

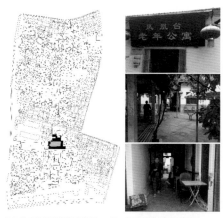

老年公寓位于荷花塘高岗里19号，是重要的文物保护单位。
The apartment for the olds is located in No.19, Gaogangli. It is one of the important legacy in Hehuatang historical area.

平面 | Plan

模型 | Model

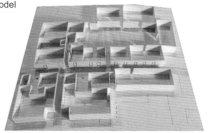

新类型 | New Type (Prototype)

新类型 | New Type

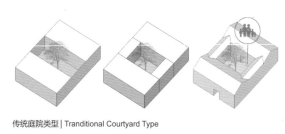

传统庭院类型 | Tranditional Courtyard Type

新类型生成 | Process of New Type

新类型发展 | New Type Development

最初类型 | Original Type

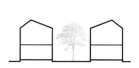

屋顶方向调整 | The Direction Adjustment of Roof

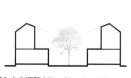

阳台方向调整 | The Direction Adjustment of Balcony

聚落布局 | Settlement Layout

类型变化 | Type Changes

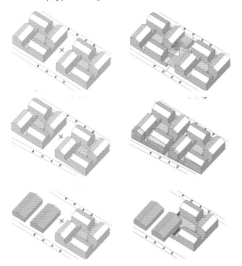

组团街巷布置 | Layout of Patterns

组团内部被设计成串联起重要保护建筑的市民花园,它像项链一般串起了老建筑与新建筑之间的对话。
In the inner part of patterns, a citizen garden is created as a necklace to connect the dialogue between historical buildings and new buildings.

总体街区布置 | Layout of Block

场地内原老年公寓旧址作为重要文化保护单位改造为荷花塘博物馆,以其为中心划分场地不同区块。
As an important cultural protection Unit, the original apartment for the olds is transformed into Hehuatang Museum, which is taken as the center to divide the site into different blocks.

建筑密度 | Building Density

既有城市肌理 | Existing Urban Fabric

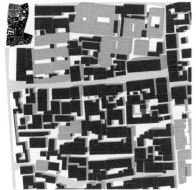

18 614 ㎡

选取场地区域进行建筑密度类比,灰色建筑为历史保护建筑。场地内老建筑多为一层,局部二层。
A area in the site is selected for density training. The grey buildings are historical heritage. Most buildings in the site are one story, few are two.

原型模式 | Prototype Pattern

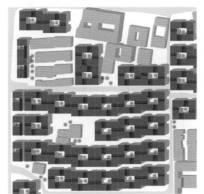

18 640 ㎡

以两层四合院住宅原型填补场地,使建筑密度保持相同。原型之间的错动形成街巷的弯曲和扭动。
The prototype of two-story siheyuan are filled in the area, keeping the densitiy same. The streets are crooked because of the range between the houses.

备选模式 | Alternative Pattern

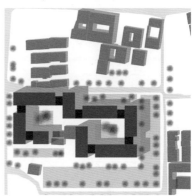

18 462 ㎡

以几幢七层高的住宅楼围合出相应尺度的内院,建筑密度保持相同,保留院与建筑的关系。
A big courtyard is enclosed by a few of seven-story buildings. The relationship between house and yard is the same while keeping the density same.

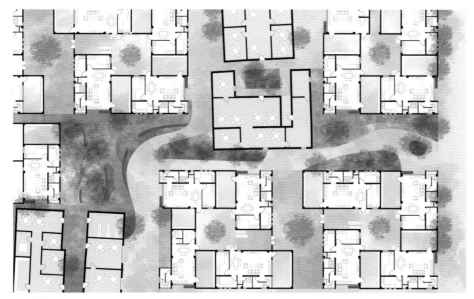

局部放大平面 | Enlarged Plan

轴测图 | Axonometry

Beaming and Connection
辐射·链

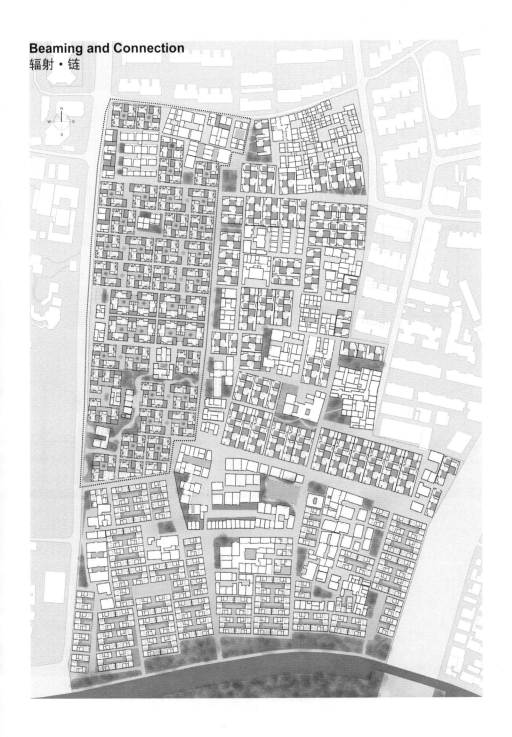

布局规则 | The Rules

1. 坐落在荷花塘19号的历史建筑改造为新的活动中心，用于展示历史建筑，也作为居民公共和休闲活动场地。
   The historical buildings located in No.19 in Hehuatang are functioned as new activity center of the area. It is used as historical building displaying and the place for residents' public and leisure arctivities.

2. 车行道将区域分成几个组团，并将周围的街道联系起来。其宽度≤7 m。
   The roads for cars divide the area into several blocks and connect the streets around. the weigth of roads ≤7m.

3. 活动中心附近的历史建筑可用于文化展览、茶室、疗养院等。
   Other historical buildings around the activity center can also be used as culture exhibition/teahouse/nursing apartment etc.

4. 组团内限于自行车和步行，宽度≤3 m。
   The inner blocks are only for bicycle and pedeistrian the weigth ≤3m.

5. 组团内历史建筑的二层也可用做借宿中心。
   Historical buildings in the blocks can also be functioned as public living center in second floor.

6. 道路两旁用于服务和商业。
   Space along the roads can be used as service or commerce.

总平面图 | Master Plan 1:500

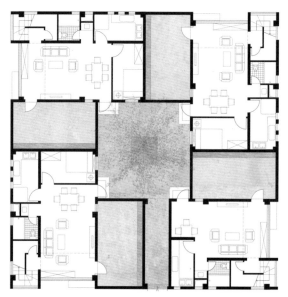

单元一层平面 | First Floor Plan of the Unit

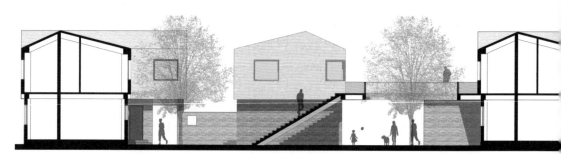

街区剖面 | Cross Section

街区立面 | Street Facade

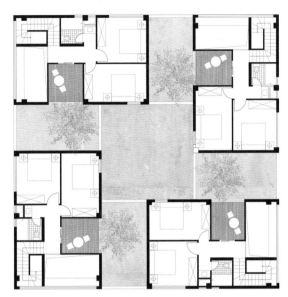

单元二层平面 | Second Floor Plan of the Unit

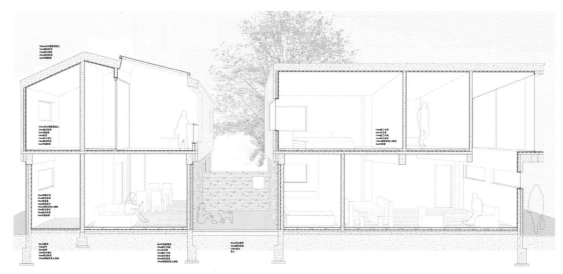

构造大样 | Construction Detail

街景 | Street View

院落场景 | Yard Views

类型调研 | Typological Survey

调研位置 | Survey Location

平面 | Plan

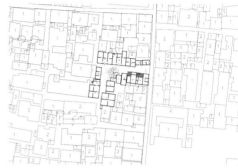

模型 | Model

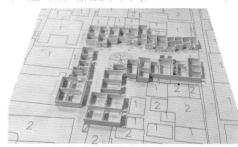

调研位置位于荷花塘北部，两边的建筑被两条走道分割，同时道路相交处有一个院子放大的节点。
The location of the research is in the north of the Hehuatang. The houses are divided by the roads, and there is a yard at the intersection.

新类型 | New Type (Prototype)

新类型 | New Type

传统庭院类型 | Traditional Courtyard Type

新类型 | New Type

新类型发展 | New Type Development

最初类型 | Original Type

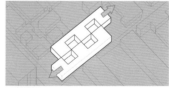

对半错位 | Dislocation

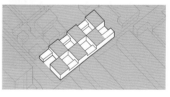

屋顶调整 | Adjustment of the Roof

聚落布局 | Settlement Layout

轴测分解 | Axonometric Decomposition

屋顶采用的是单向坡屋顶，朝向分为朝南和朝东两种。根据功能，中间四块屋顶为实，末端两块屋顶为虚。
The roof is pitched, facing north and east. And the four roofs in the middle are solid, the roofs at the end are vacant.

二层的结构采用的是传统的屋架结构，但是采用的材料是混凝土。
The second floor is traditional roof structure. The material of the structure is concrete, instead of wood.

一层的结构为砖砌体结构，同时面向院子的面为虚，面向街道的面则为实。
The structure of the first floor is brick structure. The facade facing the street is solid, while the facade facing the yard is vacant.

组团街巷布置 | Layout of Patterns

组团内部的市民花园像项链一般串起了老建筑与新建筑，为居民提供了重要的休闲活动场地。
The gardens in the group as a necklace link the new and the old architecture, providing activity places for the residents.

总体街区布置 | Layout of Block

将荷花塘内部原老年公寓旧址改造为荷花塘博物馆，以其为中心放射出的车行道划分场地。
As an important cultural protection Unit, the original apartment for the olds is transformed into Hehuatang Museum, which is taken as the center to divide the site into different blocks.

建筑密度 | Building Density

既有城市肌理 | Existing Urban Fabric

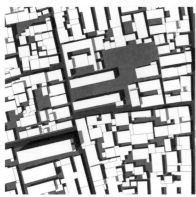

18 614 ㎡

原型模式 | Prototype Pattern

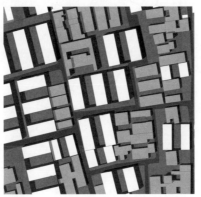

18 640 ㎡

备选模式 | Alternative Pattern

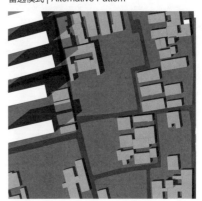

18 462 ㎡

轴测图 | Axonometry

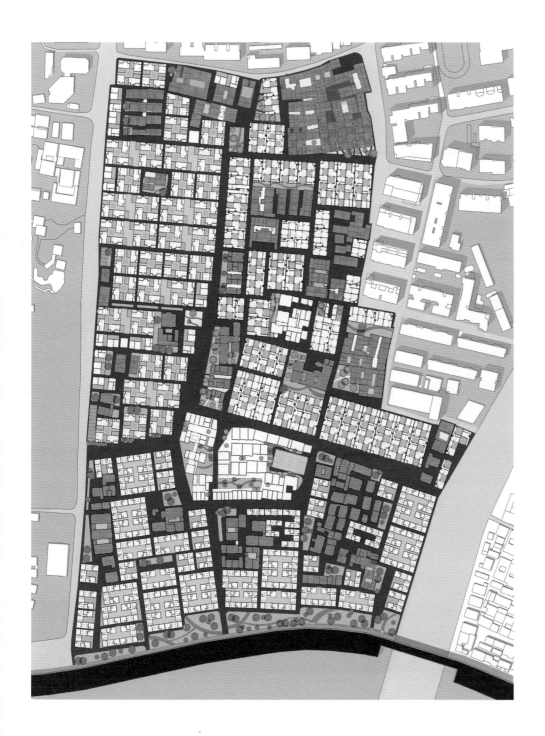

总平面图 | Master Plan

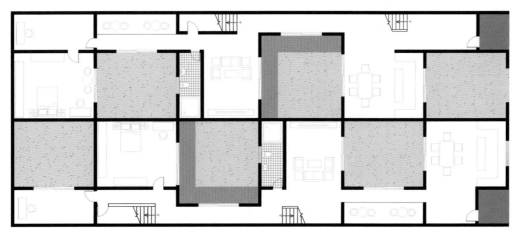

单元一层平面 | First Floor Plan of the Unit

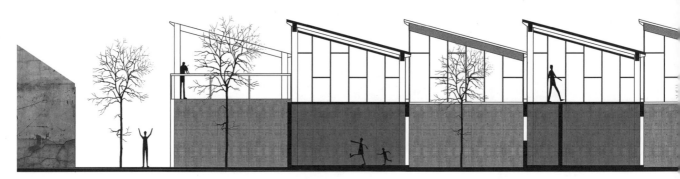

街区剖面 | Cross Section

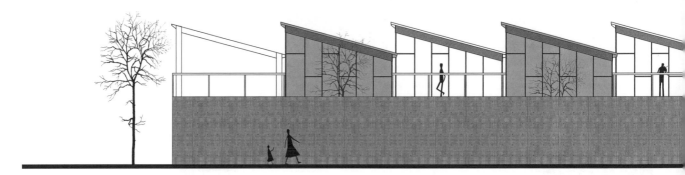

街区立面 | Street Facade

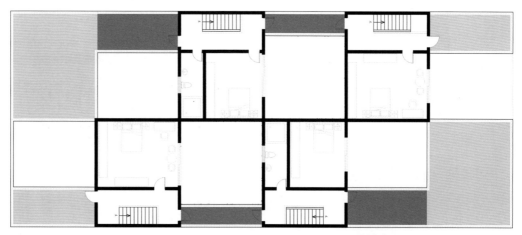

单元二层平面 | Second Floor Plan of the Unit

局部放大平面 | Enlarged Plan

场景透视 | Views

06

写作训练
Writing

概述：硕士生四篇习作

Brief Overview: Four Papers by Master Students

这四个主题的分配是在一个大的框架下进行的，就是思考城市形态学研究及其对于城市设计的价值。有一个概念性的路径：从南京的院落住宅到它随时代的变迁过程，从荷花塘道路和街道系统的阅读到附近老门东地区最近十年间更新设计的结果。这些论文完成得很好，并通过草图和图解进行了充分的解释和说明。

丁园白和王安安给出了一个很好的报告，以了解南京普通（或"普遍"）院落住宅的发展过程。

吴舒和范琳琳基于功能进行了很多说明，但最终他们关于形式的推理比功能问题的研究更加重要。

施剑波、吴则鸣的工作非常好地对事实进行了有趣的描述，并绘制了街道的剖面图。

今天从夫子庙漫步到老门东，再到荷花塘，会是一段穿越不同城市更新思想的奇异旅程。韩珂与刘巧的论文尝试以一种简单而清晰地方式来阐述其背景。

The assignment of those four topics has been made in the framework of a reflection about urban morphology and the worth of such kind of research for urban design. There is a conceptual path: from the Nanjing courtyard-house to its change in times, from the reading of the system of roads and streets in Hehuatang to the outcomes of the last decade regeneration design in the near district of Laomendong. The papers are well done, sufficiently informed and fully illustrated by sketches and schemes(diagram).

Yuanbai DING and An'an WANG gave a good report to understand how the general (or "universal") development of the courtyard house type was working in Nanjing.

Shu WU and Linlin FAN wrote a lot about functions, but at the end their reasoning about shapes is more important than the study on questions of functions.

The one by Jianbo SHI and Zeming WU is a good job, with an interesting description of the reality of facts, and also drew sections of the streets.

Taking a walk from Fuzimiao to Laomendong until Hehuatang is today an extraordinary journey crossing different ideas of urban regeneration, the Ke HAN's and Qiao LIU's paper tries to describe the background of that in a simple and very clear way.

南京地区院落建筑的类型学研究——以荷花塘为例
The Typological Research of the Courtyard Building in Nanjing-Hehuatang for Example

丁园白　王安安
Yuanbai DING　An'an WANG

在南京荷花塘地区留存的现有居住建筑依然保留着明清时期典型的传统居住空间模式。同时也暴露出这种模式与当前的生活方式和社会结构碰撞后出现的一些问题。因此，我们尝试从荷花塘地区的传统院落中提取六种基本类型，诠释传统居住空间的空间特征。通过对比分析中国的不同时期、不同地域的庭院空间，试图找到新时期城市建设与传统居住模式相结合的可能性。

南京是六朝古都，是国家历史文化名城，拥有近2 500年的城建史，累计约450年的建都史，沉淀了丰富的历史文化遗产，形成了独特的人文景观，在中国乃至世界都城建设史上具有重要地位。城南历史城区是南京历史文化名城价值与特色的重要载体，是明清南京历史风貌的典型代表地区，其近年来逐渐成为学术界、各级政府与社会媒体所关注的焦点。

在众多历史建筑中，最能反映丰富市井生活的当属居住建筑。而荷花塘地区作为南京地区反映城南居住地区、居住功能最完整的区域，保存了最完整的明清住宅区、最清晰的历史格局、最完整的传统风貌。因此本文以荷花塘街区内现存典型院落为研究对象，应用类型学的方法将其空间组织抽象归类，对中国南京地区典型的院落空间布局以及空间构成、空间特质进行尝试性解读，并在此基础上，分析其反映出的当地人的居住模式与社会形态。

The existing residential buildings in Hehuatang area of Nanjing retains the typical traditional residential space patterns in the Ming and Qing dynasties. But it also exposed some problems facing the present lifestyle and social structure. Therefore, we use the method of typology to extract six basic types from the traditional courtyards of Hehuatang area and try to interpret the spatial characteristics of traditional residential space. And then we make an analysis and comparison of the courtyard spaces in different periods and regions. In the end, we try to find the possibility of combining the urban construction with the traditional residential mode in the new era.

Nanjing, a national historical and cultural city, is the capital of six dynasties in Chinese history. Nearly 2 500 years been urban and a total of about 450 years been the capital. Accumulated rich historical and cultural heritages and formed a unique cultural landscape, Nanjing has an important position in the history of the capital construction in China and even in the world. The south city is an important carrier of the history and culture of Nanjing .It is a typical representative of the Ming and Qing dynasties of Nanjing historic area, which in recent years has gradually become a focus of academic circles, government and social media at all levels of attention.

In many historical buildings, which can reflect the abundant life of people is the residential one. As the representative of residential function in south Nanjing, Hehuatang area has the most complete preservation of the residential buildings of the Ming and Qing dynasties .It has the clearest historical pattern and the most complete traditional style. Therefore, we concentrate on the existing typical courtyard in the Hehuatang area, and use the method of typology to classify the spatial organization of each courtyard, and try to read the distribution and space characteristics of typical courtyard. And on this basis, we can analyze the mode of living and social forms of local people.

1 类型学的相关理论与方法

类型学是一种分组归类方法的体系，是一种具有深厚的哲学基础与多样的方法体系。在18世纪，分类处理，系统连续、统一的方法被应用于建筑，因而有了建筑类型学。建筑类型来源于传统中的建筑形式，通过研究这种具体形式的外在结构而分析出一种内在深层结构。尽管外在的具体形式会受到各种因素的影响，但这些建筑形式都能被抽象成同一种建筑原型。同一类型的不同建筑之间也互为变体。

南京荷花塘传统民居院落空间层次丰富，保存完好，形式灵活多变。从类型学的角度，我们可以将传统院落的平面形态进行分类对比，总结出几种基本原型。通过类型分析，我们清晰地获得了南京荷花塘地区传统民居的总体形态特征。

2 荷花塘地区传统院落居住空间类型研究

（1）"一"形院落

"一"形院落是一面建筑三面围墙的院落，规模较小，一般供一户居民使用，也有的是群体院落空间的单体。"一"形院落是最简单的院落类型，其形成分两种，一种是院落与建筑同时建成，另一种是随着家庭结构改变或所需居住面积的增加，在原先单体建筑上加建庭院所得。院落与建筑之间一般设有一条檐廊作为过渡空间。建筑入口分为两类，一类从院落门中进入，通过庭院空间进入室内，此类形式院落公共性较强。另一类是以建筑作为入口，庭院作为建筑后院使用，此类院落私密性较强。建筑房间布局一般顺着院落走向划分，朝向相似，无明显等级区分。建筑一般由一家人共同使用，通过房间划分面积的大小来确定主要空间与辅助空间。由于院落三面是围墙，与自然接触面积大，能很好地与周围环境结合。院落承载的是家庭内部活动，较少能促进邻里交流。

1 Related theories and methods of typology

Typology is a system method of classification. It has a profound philosophical basis and diversity system. In eighteenth century, a continuous and unified method of classification processing system used in construction and it called architectural typology. Types are derived from the traditional architectural form, through the external structure of this concrete form to analyze a deep relations. Although the specific form of the external will be affected by various factors, but all these forms can abstracted into the same architectural type. Different buildings of the same type are also variant.

Traditional courtyard spaces in Hehuatang area are abundant and well preserved. They have flexible forms. In the typological point of view, we can compare the plane shape of traditional courtyard and summarize several basic types. Through the analysis of these type, we can clearly get the overall characteristics of traditional residential buildings in this area.

2 The traditional courtyard types of Hehuatang area

(1) " 一 " courtyard

This type of courtyard has one side of building and three side of walls.It is small and generally for a family,but also for some monomer groups.It is the simplest courtyard type.The formation is divided into two kinds: one is the courtyard and the building built at the same time;the other is with the changes of family structure or the needed for more space,the courtyard was built after the original single building.Between the courtyard and the building is generally equipped with a gallery as a transitional space.The building's entrance is divided into two categories.One is from the door on the walls.You must enter the courtyard space and then into the room.This kind of courtyard is more public.Another is enter from the building.The courtyard is as the backyard.This courtyard is much privacy.The general layout of rooms is along the courtyard and divide the building into smaller.The orientation is similar and there are no obvious distinction between levels.Because there are three courtyard walls,and the area contacting the nature is ample,it has an excellent combination with the surrounding environment.The courtyard is for the family activities and have no contribution with neighborhood communication.

（2）"L"形院落

"L"形院落是两面建筑、两面围墙的组合方式，两面建筑垂直相交。一般面南建筑是主要空间，东向或者西向建筑是辅助空间。"L"字形院落有两种形成规律。其一是"L"形建筑是一个独立的单元。其二，院落起始并不是"L"原形出现，而是"二"形院落，随着时间的推移、人们生活生产方式的改变，对院落空间产生了新的要求，而社会结构的变化也带来院落使用方式的变化，最初的院落原型通过加建演变成"L"形的院落单元，也可能是"U"形单元由于院落空间的改变，经过人为地拆除后形成"L"形院落。"L"形院落的檐廊也设在建筑与庭院交接处，根据东西向空间使用性质的不同，有的东西向与建筑交接处设有檐廊，有的不设。相比于"一"形院落，"L"形院落具有一定的围合度和内聚性，也与自然有较好的联系。

（3）"U"形院落

"U"形院落的形成是单元原型的直接出现过程。可理解为该院落原型自身的特性规律。单元原型院落本身对外开放了一个方向。这类院落或直接对外衔接历史街区街巷，或直接面向内部街巷，满足各种层级逻辑上的对外衔接要求。

(2) "L" courtyard

"L" courtyard is a combination of two sides of the wall and two vertical sides of the building. Generally the south building is the main space and the east or west side is the auxiliary space."L" courtyard has two kinds of forming rules, one is the "L" building is an independent unit. Second, the courtyard is not starting with a "L" prototype. It is from the " 二 " prototype courtyard. As the time goes by, people's living habit and production ways changed, there are new requirements for courtyard space, or the social structure's development changes the way of using the courtyard. The original prototype construction developed into "L" courtyard by adding a building .It may also be due to the changes of courtyard space from the "U" courtyard and artificially removed to the formation of "L" courtyard. The gallery of "L" courtyard is also located between the building and the courtyard. Depending on the different of use property of the east and west space, it is maybe located with the gallery in the junction of the building and the eastern or western courtyard. Compared to " 一 " courtyard, "L" shaped courtyard is more enclosure and cohesion. It have a good relationship with nature.

(3) "U" courtyard

The formation of such courtyards is the direct emergence of the prototype unit process. It can be understood as the courtyard prototype itself characteristic rule. Unit prototype courtyard itself open to the outside world in a direction. Such courtyard is directly outside the historical block street, or directly external to the internal streets, to meet the various levels of logical external convergence requirements.

（4）"二"形院落

"二"形院落代表着南京历史街区中绝大部分的具体院落形态类型，是南京历史街区院落形态的主导角色，其形成规律在一定程度上可以阐述南京历史街区院落形成的主导规律。院落往往从一个原型单元开始，以一个方向为院落生长的主导轴线，原型单元沿着主导轴线重复。这一过程中，原型单元重复的数量或多或少，体量或大或小，但始终沿着其主导轴线方向发展，进行着同一类型化原型单元的重叠复制，主导轴线方向往往形成较为强烈的序列感。同时，这类院落最终呈现出强烈的传统等级观念。沿着主导轴线方向的序列重复越多，呈现出的院落序列感越强烈，院落的进深越大，院落的空间感受越庄重神秘，越吸引人走进院落一探究竟。

（5）"回"形院落

"回"形院落是以独立单元原型出现，院落保持着自己完整的院落形态，四个方向均没有对外界开口。这并不代表着这类院落不需要自然环境及与外界自然的衔接和交融。在院落的中央或者稍偏的位置开放院落，形成院落内部独立的天井或更大庭院，每一个单元院落原型形成属于自己的小宇宙。以独特的方式和自然进行着对话，真正做到与自然之间的张弛有度。

(4) "=" courtyard

These courtyards represent the vast majority of concrete courtyards in the Nanjing historic district. They are the leading role of the courtyards in the Nanjing historic district. To some extent, the formation rules of the courtyards can be used to illustrate the dominant laws of the Nanjing historic districts. Courtyards often start with a prototype unit, in a direction for the courtyard growth of the leading axis. The prototype unit along the dominant axis repeating. In this process, the prototype unit duplicates the number of more or less, large or small, but is always is along the direction of its dominant axis. The same type of prototype unit overlaps replication, the dominant axis direction often forms a more intense sense of sequence. At the same time, these courtyards eventually show a strong traditional hierarchical concept. The more the sequence along the dominant axis is repetitive, the more the sense of courtyard sequence to be represented is intense, the deeper the depth of the courtyard is, the more the space perception of the courtyard is solemn and mystery, the more to attract people into the courtyard to find out.

(5) " 回 " courtyard

These courtyards is an independent unit prototype form. The courtyard maintains its own courtyard form that the four directions are not open to the outside world. It does not mean that such courtyards do not need the natural environment and the connection and intergration with external natural world. In the courtyard of the central or less open position of the courtyard, the independent courtyard or larger courtyard is formed within the courtyard, so that each unit yard prototype generates their own small universe. It has a dialogue with nature in a unique way, and truly achieves a degree of relaxation with nature.

3 南京与其他地区民居的差异性比较

荷花塘街区内的多样传统院落是南京地区民居的重要组成部分，在江南的民居中具有一定的代表性。整体来看，民居的院落组织受自然条件的影响较大，气候、日照、降水等与居民生活相关的因素都会反映到建筑与院落的形体布局上去。

北方民居如北京四合院、山西窑洞的合院式布局，房屋间以廊相连或各自分离，院落私密性强而尺度开敞近人，适合在院内进行室外活动；同属南方地区的徽州民居的厅井式布局，房屋间相连而院落窄小似井，仅可作通风采光之用而不适宜人的户外活动。而江南一带民居，以南京荷花塘街区为例，可以看出其院落与房屋的比例关系介于上述二者之间，对于居住者的活动导向与限制并不明确，存在室内室外之间的多种可能性。

4 南京地区传统院落住宅空间的更新

随着经济社会的飞速发展，南京老城南许多未曾得到严格保护的传统院落住宅正在发生形态的变化，常见的有两种方式：一种会导致院落空间形式发生改变，诸如置换或重新划分空间的使用功能、改建已有房屋、在剩余的院落空地搭建新房等等；另一种则会破坏合院沿轴线叠加所产生的序列感，诸如拆除墙体、打开封闭院落的行为使得院与房屋的纵深关系弱化，而仅仅强调了房屋间横纵方向的不同组合方式。

3 Comparison of the Differences between Houses in Nanjing and Other Places

The diverse traditional courtyard in Hehuatang area is an important part of residential houses in Nanjing, which is a representative sample of residential houses in Jiangnan region. As a whole, the organization of residential courtyard is mainly affected by the natural conditions. The factors which are closely related to residential life, like climate, sunshine duration and precipitation, will be directly reflected in the shape and arrangement of the building and courtyard.

The courtyard-style arrangement of northern houses, such as the Beijing Siheyuan, Shanxi cave dwellings, the buildings are separated from each other or connected with corridors, and the courtyard is private and close to the scale of people, suitable for outdoor activities in it. Also with the patio-style arrangement of southern houses, such as Huizhou residential houses, buildings are connected to each other and the narrow courtyard is like a well, which is only for ventilation and lighting and not suitable for people's outdoor activities. Taking Hehuatang area in Nanjing as an example of the residential areas along the Yangtze River, we can find that the proportional relationship between courtyard and buildings is between the above two types. So the houses in Hehuatang area are not clearly activity-oriented and restricted for the occupants. There are various possibilities for people to choose indoor or outdoor activities.

4 Renewal of Residential Space with Traditional Courtyard in Nanjing

With the rapid development of economy and society, many of the traditional courtyard houses in the old South City of Nanjing which have not been strictly protected are undergoing changes in shape. There are two common ways of the changes: One will lead to changes in the form of courtyard space, such as replacement or reuse of space function, rebuilding existing houses, building new houses in the remaining courtyards and so on; The other will destroy the sequence of the houses along the axis generated by stacking buildings and courtyards, such as the removal of the wall, opening the closed courtyard. Those behaviors make the relationship between courtyard and building weakened, only emphasize the different combinations of the vertical and horizontal buildings.

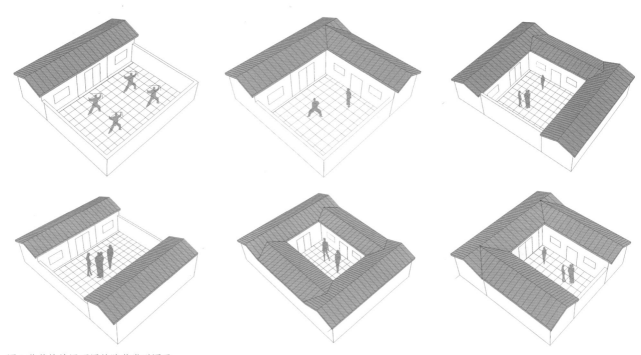

图1 荷花塘地区不同的院落类型图示
Figure 1 Diagram of different courtyard type in Hehuatang area

而造成这种现象的原因是多样的，关键的因素有：一是随着时间的发展，同一地块的居住密度在持续增加。原属于一户的空间现居若干户，且均有对于完整的生存起居空间的需求。那么在空间形式与面积不满足的情况下，只能在自身空间范围内进行相应改造，从而导致院落住宅的外壳下面包裹着种类与数量繁多的生活空间。二是现代生活方式与传统居住空间的矛盾，使得旧有的空间组织方式已不能很好地适应现代生活的需求，居民多以家庭为单位对建筑进行了改造，从而破坏了院落组团乃至整个街坊的肌理。三是对于传统院落住宅的管理与维修不当，以及居民缺乏古建保护意识。

Reasons for the phenomenon are diverse, but the key factors are: Firstly, as time goes on, residential density of the same plot continues to increase. A house which was originally belonged to one family is divided to several households now, and each of them needs the full set of living space for their daily life. Under the condition that form of space and the dwelling area are not satisfied with people's needs, people can only make the corresponding transformation in their own space. As a result, different types and a large number of living space are wrapped under the shell of courtyard houses. Secondly, modern living styles and traditional living space are contradicted to each other, so that the old way of space organization cannot be well adapted to the needs of modern life, and many of the residents transform the buildings in the unit of family, thereby undermining the texture of the group of courtyard and the entire neighborhood. The third is the improper management and maintenance of traditional courtyard house, as well as the lack of residents' awareness of the ancient building protection.

现在对于传统院落住宅类型的总结是在为老城的更新打下基础，以在不失去传统住宅特色的前提下实现对历史街区、地段的保护，同时激活该地区人们的生活。

5　结语

对比现今房地产开发的住宅小区，人们的生活与地面脱离而择于高处，邻里间的交流也仅限于入户时相遇的一声招呼，在住家的社交圈单一而封闭。这些现状到底是在新时代发展中人们对于生活方式的自主选择，还是在适应新型居住单元过程中的被动接受呢？

而院落作为古代南京城市居住空间组织的重要原型，在当代城市更新的建筑实践中依然处于重要的地位。对于南京荷花塘地区院落类型的研究是在给现在的居住模式提供一种改变的可能性。发现原有院落类型在现代居住生活中的优势并将其保留下来，建设还原旧时建筑院落的组织关系和居住体验的住宅，同时满足现代人的生活需求。可能会以一种新型的低层高密度住宅的方式呈现在人们面前。

Now, the conclusion of traditional courtyard-style is to lay the foundation of renewing the old city, realize the protection of historic blocks without losing traditional residential features, and activate the life in the region.

5　Conclusion

Considering the current estate development of residential districts, people live in the tall buildings far away from the ground, and the exchange between the neighbors is only limited to greeting each other when entering their houses. The social circle of people living around is single and closed. A question we need to think about is that the current situation is people's choice for the way of life in the developing new era, or the passive acceptance in the process of adapting to the new living unit?

The courtyard, as an important prototype of ancient Nanjing city living space organization, is still placed in a significant position in the practice of modern urban renewal. The study of the type of courtyards in Hehuatang area of Nanjing provides the possibility of a change in the present mode of living. Find the advantages of the original courtyard type in modern living and keep them, so that we can make the construction of houses that keep the organization of old buildings and living experience of residential, and meet the needs of modern life at the same time. Maybe a new type of low-rise high-density residential houses will be presented in front of people.

参考文献

1　卢波．类型学视野下的苏州传统院落空间形态特征探究．南方建筑，2013(4)：91-95.
2　吴超．南京老城南门东历史街区传统院落布局特征．城市建筑，2013(4)：3-3.
3　张春霞，谢金之．南京荷花塘传统院落的空间形态探析．艺术探索，2015,29(2)：101-103.
4　许伯晗．南京门西传统院宅空间类型及建造特征浅析．南京：南京大学，2016.

传统院落的蜕变——类型如何随时间变化，未来该如何发展
The Metamorphosis of Traditional Courtyard - How a Type Has Changed with Time, How Can it Develop in Future

吴舒　范琳琳
Shu WU　Linlin FAN

荷花塘是南京老城南地区典型的传统民居聚集地。随着时代发展，传统院宅功能屡经更新。尤其是近代以来，城市迅猛发展，原本只满足基本生活需求的传统院宅形制发生了很大改变，传统的建筑形制也日渐消亡。因此，研究院落格局的演变过程对于理解老城南历史风貌和地段更新具有重要意义。本文分为两部分对院落形态的演变过程进行研究。第一部分以位于荷花塘地区北侧的刘芝田故居为例，研究其在不同时期院落格局的改变；第二部分介绍了近年来传统院落空间发展变化的新情况，并且对传统院落空间未来发展的前景做出预期。

1 类型如何随时间变化

（1）荷花塘民居基本类型

对荷花塘民居院落类型进行提取，可以归纳出五种基本单元形式：分别是"一"形、"L"形、"凹"形、"C"形和"回"形。荷花塘整体建筑群基本由这几种形式组合而来。五种基本形态，可以帮我们定义最初的平面。

Hehuatang area is a typical traditional residence in the Old South City of Nanjing. As the development of times, functions and types of courtyard houses keep changing. Especially in recent years, because of the rapid development of city, traditional courtyard houses can hardly meet the basic needs of modern life. Old courtyard houses and living memories are disappearing.Therefore, the study of the evolution of the courtyard pattern is of great significance to study the historical features and location update of the Old South City. This paper is divided into two parts to study the evolution process of courtyard form. The first part takes the former residence of Zhitian LIU as a typical example, which is in the north part of Hehuatang area, and focuses on the change of courtyard pattern during different periods. The second part of this paper introduces the traditional courtyard space development and change in recent years, and make expectations of the future development of traditional courtyard space.

1 How a type has changed with time

(1) Basic types of residence in Hehuatang

The courtyard types of residence in Hehuatang can be induced into five basic unit forms: "一" type, "L" type "凹" type, "C" type and "回" type. In Hehuatang area, we can find that traditional houses with a series of courtyards are always combined by several forms of this types. These five basic types will help us to identify the original plane of traditional courtyard houses.

（2）清末民初时期

刘芝田故居始建于清末，从原始平面图上，可以看出其原始院落组合的空间关系。虽然故居群不同于中国传统建筑群组织严格的对称手法，但依然遵循沿轴线展开空间序列的方式。刘宅被内部两条巷弄分为三部分，中间部分是整组住宅的主要空间。主入口位于建筑群东北角，进入院宅后穿过东北侧的两进院落（即轿厅）进入巷弄，此处是整组院宅轴线转折之处；穿过巷弄继续向南走，进入中间部分的院落后轴线转为东西向；或在巷弄处西转，沿巷弄走，依次进入各组院落。

(2) The late Qing dynasty and early Republican period

The former residence of Zhitian LIU was founded in the late Qing dynasty. From the original plan, it can be seen that the space relationship of the original courtyard combination. Although the former residence group is different from the strict symmetry technique that traditional Chinese architecture group organized ,it still follow the way that expanded the space sequence along the axis. The former residence of Zhitian LIU was divided into three parts by the internal two lanes, the middle part is the main space of the whole group.The main entrance is located in the northeast corner of the building group. Entering the courtyard house, passing through two courtyards by the northeast side and geting into the alley, here is the turning point of the whole set of home axis: through the alley to the south, into the middle part of the courtyard, the axis turns to the east and the west. Or turn west after the alley, walk along the lane, enter the courtyard in turn.

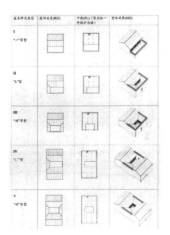

图1 荷花塘院落基本类型
Figure 1 The basic type of Hehuatang courtyard

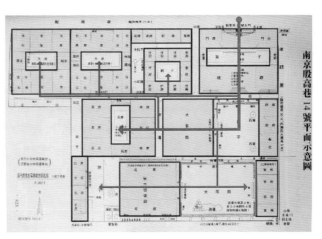

图2 刘芝田故居清末原始平面图
Figure 2 The plan of the former residence of Zhitian LIU in the late Qing dynasty
来源： 摄于刘芝田后人故居内

（3）社会主义早期

随着后人的陆续搬离，整组建筑群逐渐衰落，最南侧建筑群消失，北部的一个建筑变为他用。为了与现有道路连接，主入口的位置移动到整组建筑群东侧的孝顺里上，因此整体建筑的轴线序列也不复存在。一条窄巷成为联系各院落建筑的主要通道。而住户仍属一个大家庭，因此单组建筑内部空间格局并没有太大的变化。

(3) Early socialist period

As some future generations of the big family moved away from the buildings, the whole group of buildings gradually declined. The south part of the houses was disappeared, and a house in the north part was changed for other use. To connect with the existing road Xiaoshunli on the eastern side, the main entrance of the whole group of buildings moved to the eastern side. So the original axis didn't exist anymore. The narrow lane became the main road which connected each courtyard houses. However, people who use the houses still belong to a family, so internal spatial pattern of the houses didn't change much.

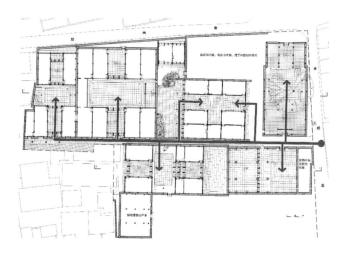

图3 刘芝田故居总平面
Figure 3 The master plan of the former residence of Zhitian LIU

图4 刘芝田故居社会主义早期平面图
Figure 4 The plan of the residence of Zhitian LIU in early socialist period

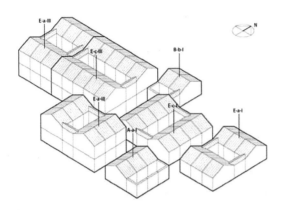

图5 刘芝田故居社会主义早期变迁
Figure 5 The development of the former residence of Zhitian LIU in early socialist period

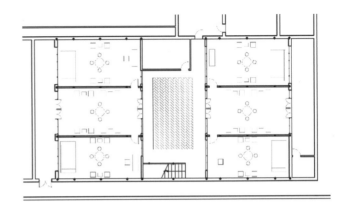

图6 刘芝田故居 14-4 号住宅近现代测绘平面
Figure 6 The modern surveying plan of the former residence of Zhitian LIU number 14-4

（4）近现代时期

经历产权分割后，刘芝田故居的产权发生改变，不再属于一户家庭。建筑群的整体性减弱，轴线序列消失。如今产权所有者以"间"为单位供外来人员租住。为满足现代生活需要，房屋使用者进行了不同程度的改扩建，原先明晰的单元类型变得模糊，转变为一种适应现代生活需要的新类型。然而新类型并不具备典型性，相反，它抹去了荷花塘地区原有建筑的风貌。以刘芝田故居为例进行研究，可以发现现在每户家庭住一间房。原本独立功能的"一间"房间内部被再划分，加入厨房、卧室、卫生间等复合功能，院子面积也由于加建建筑而减小。

(4) Contemporary period

After the division of property rights, the property right of the former residence of Zhitian LIU had changed. It no longer belong to one family. The axis completely disappeared, and each courtyard house of the building group became an independent one. Today, property owners rent the rooms to outsiders. To meet the needs of modern life, people living in houses have added some buildings to the houses. Therefore, the original clear cell types became blurred, even changed into a new type which adapt to modern life. However, the new types are not of typicality. In contrast, it erased the original architectural style in Hehuatang area. I choose one courtyard house of the former residence of Zhitian LIU as an example to study. We can see that each family live in one room nowadays. This one room of original independent function was divided into the kitchen, bedroom, bathroom and other complex functions. The yards area are reduced due to the construction of the building.

2 未来该如何发展

在传统院落中，庭院是院落空间的核心，但在现状中这种传统形式渐渐消失。庭院原有的通达性由于多户居住模式的引入而受阻，变得内向而封闭。庭院原本是公共空间，但是由于多户居住，服务空间不足，造成了庭院被瓜分，沦为服务空间。希望在未来院落改造中，能够打破庭院现在的封闭模式，恢复庭院作为组织空间的核心地位，激活庭院公共空间的性质。

以刘芝田故居14-4号住宅为例，从公共空间、入户流线、服务与被服务空间三个方面对复原平面和现状测绘平面进行分析，试图找到未来改造的方向。

（1）公共空间

由图7、图8对比可以看出，在原来的传统院落中，庭院与周边穿越性房间一起起到了公共空间的作用，由庭院到私人房间有公共空间、半公共空间和私密空间三个层次，并且由一条轴线控制。庭院之间有交叠的关系，之间由穿越性空间串联。但是在现状平面中，轴线和穿越性的灰空间都消失了，由庭院到房间只是由公共空间到私密空间，中间没有过渡。而二层的半公共空间完全被占用为服务性空间，失去了庭院作为公共交流空间的性质，杂乱无章。

2 How can it develop in future

In the traditional courtyard, the courtyard is the core of the courtyard space, but in the current situation, the traditional form is gradually lost. Because the increase of population, the original accessibility of the courtyard is lost, and the courtyard become introverted and closed. The courtyard was originally supposed to be a public space, but in the current situation, service space are insufficiency, so the courtyard are divided up, become service space sadly. In the future transformation of the courtyard, I hope that the closed pattern of courtyard can be broken, the courtyard can restore as the core position of organization space, and the public space quality of the courtyard can be activated.

Take the former residence of Zhitian LIU number 14-4 as an example to analyse the restoration plane and the present surveying plan. This analysis discussed in three aspects: the public space, the entrance streamline, the service and served space. Then we try to find the direction of the development of the courtyard.

(1) The public space

From the contrast of figure7 and figure8, it can be seen that, in the original traditional courtyard, the courtyard and the passage room surrounded together play a role of the public space. From the courtyard to the private room you should through third level of space: public space, half-public space, private space, and an axis control the whole space structure. There are overlap between two courtyard—the passage space. But in the present plane, the axis and the passage gray space are all disappeared. There is no transition space from the courtyard to the room, but only the public space and the private space. The half-public space in the second floor are used for the service function completely. The courtyard are all in a mess without the charactor as public communication space.

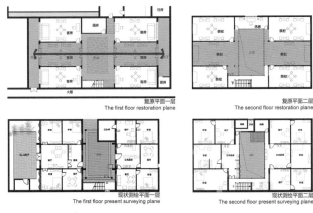
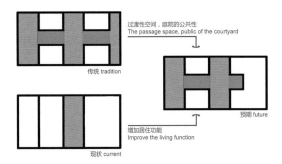

图 7 传统与现状院落中的公共空间
Figure 7 The public space in traditional and present courtyard

图 8 院落中公共空间的未来预期
Figure 8 The future of public space in courtyard

在未来的发展改造中，希望能够恢复传统形制，保持空间由公共空间到半公共空间再到私密空间的层次，重新置入穿越性质的灰空间，保证庭院空间的公共性。但是由于现代人口格局与传统有相当大的区别，也需要对传统院落空间的居住功能进行完善，保证院落能够继续容纳现在数量的人口。

（2）入户流线

在现状平面中，一层平面流线围绕庭院展开，庭院连通两个入口，住户从主入口进入庭院，从庭院进入公共厨房，再进入各自家中。二层从楼梯上来，围绕庭院形成"U"形流线。住户首先进入公共厨房区域，再进入各自家中。进而在原来的平面中，流线是由轴线串联，由火巷进入夹道（入口庭院）穿过客厅再进入中心庭院，进而进入其余房间。由楼梯到达二层，流线围绕庭院一周，进入过渡空间再进入房间。

In the development and transformation of the future, I hope that the traditional form can be recovered, ensure the three levels: public space, half-public space, private space. The passage gray space can be recovered to ensure the feature of public of the courtyard. Though because the modern population pattern and the traditional population pattern have a quite big difference, it also need to improve the living function of the traditional courtyard space, ensure that the courtyard house can continue to accommodate the current population.

(2) Entrance streamline

In the current plane, the streamline of the first plan is taken the courtyard as the core, the courtyard connects two entrances. The household enter the courtyard from the main entrance, then from the courtyard into the public kitchen, and then into their homes. The second floor come up from the stairs, around the courtyard the streamline shaped U. The household come into the public kitchen area firstly, and then into their homes. In the original plane, the streamline is connected in series with the axis, by the huoxiang into jiadao(entrance courtyard), through the living room, then into the center of the courtyard and the rest room. From the stairs to the two floor, the streamline makes a circuit around the yard, then into transitional space and rooms.

在未来的发展改造中，希望恢复入口庭院的形制，然后从入口庭院经过过渡性空间进入中心庭院，再由中心庭院作为节点组织交通，产生空间的序列感。到达二层后，希望公共流线通过庭院组织，再经由过渡空间入户。但是还要考虑安排多户生活，楼梯和入户流线要尽量具有通达性，空间组织要尽量紧凑简洁。

In the development and transformation of the future, I hope that the pattern of the entrance courtyard can be recovered, then from the entrance courtyard to the central courtyard passed by the transitional space, taking the central courtyard as a node to organize traffic, to get the feel of sequence sense. After reaching the second floor, it hopes that the public streamline organizes by the courtyard via the transitional space then into the rooms. Though it also need to consider the arrangement of multi family life, the stairs and entrance streamline must be accessibility, spatial organization should be compact and concise.

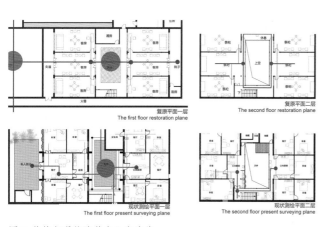

图 9 传统与现状院落中入户流线
Figure 9 The entrance streamline in traditional and present courtyard

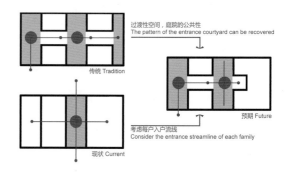

图 10 院落中入户流线的未来预期
Figure 10 The future of entrance streamline in courtyard

（3）服务与被服务空间

(3) The service and serve space

图 11、图 12 中红色区域表示服务功能空间，绿色表示被服务功能空间。在现状中，服务功能分布混乱，服务功能空间上下不对位。现状一层中，庭院的两侧成为公共厨房和储藏杂物的地方，压缩了整个庭院的使用面积。现状二层中，服务空间占用走道部分，多为住户的厨房和储藏杂物的地方，庭院更甚，成为住户晾晒衣物排出污水的场地。总体而言，现状的庭院因为被服务空间占据变得十分消极。而从原来的平面可以看出，服务功能空间与庭院是有隔离的，庭院作为被服务功能空间出现，比较宜人，尤其是在二层，杂物都被堆放在房间内，过渡空间和庭院空间十分完整。

Red area in the figure11 and figure12 is means the service space, green area is means served space. In the present situation, the service are arranged in a random, the service space is not alignment in the vertical direction. On the first floor, the space beside courtyard become public kitchen and storage areas, the use of the entire courtyard area is compressed. On the second floor, the aisle is occupied by the service space in which mostly are kitchens and storage areas, the worse situation has occured in the courtyard where the residents use them as clothing drying and waste water discharging space. Overall, the current situation of the courtyard is very negative because of be occupied by service space. The original plane shows that the service space is seperated from the courtyard which is pleasant as an served space. Especially in the second floor, things is kept in the room that makes the transitional space and the courtyard space integrity.

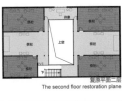

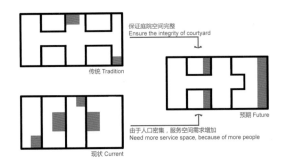

图 11 传统与现状院落中服被空间
Figure 11 The service and served space in traditional and present courtyard

图 12 院落中服被空间的未来预期
Figure 12 The future of the service and served space in courtyard

在未来的发展改造中,希望把服务性质空间放置在角落,不占用庭院和过渡空间,保证庭院的完整性。同时服务性质空间上下对位,方便使用。

In the future transformation, we hope to put the service space in the corner so that not to occupy the courtyard and the transitional space and ensure the integrity of the courtyard. At the same time, the service space is alignment in the vertical direction and easy to use.

3 结论

我们选取了14-4号院落,从公共空间、入户流线、服务与被服务空间三个方面入手,分析了复原平面和现状平面中庭院的类型,进而对未来可能的发展方向做出预期。公共空间方面,希望能够恢复传统形制,重新置入穿越性质的灰空间,保证庭院空间的公共性。入户流线方面,希望保持原来空间序列的起承转合,也要结合实际户型做出调整。服务与被服务空间方面,希望把服务性质空间放置在角落,不占用庭院和过渡空间,保证庭院的完整性。同时服务性质空间上下对位,方便使用。希望能够举一反三,为荷花塘院落之后的发展提供借鉴。

参考文献

汤建华. 老城民宅生长性策略——南京门西高岗里21号地块更新设计. 南京：南京大学, 2016.

3 Conclusion:

We selected No.14-4 courtyard and analyzed the types of the restoration plane and the present plane from three aspects: the public space, the entrance streamline and the service and served space, then we try to find the direction of the development of the courtyard. In the aspect of the public space, we hope to restore the traditional shape and imbed the gray space which used as passageways to ensure the publicity of the courtyard. In the aspect of entrance streamline, we hope to keep the rhythm of the original space sequence and make adjustments according to the housing type. In the aspect of service and served space, we hope to put the service space in the corner so that not to occupy the courtyard and the transitional space and ensure the integrity of the courtyard. At the same time, the service space is alignment in the vertical direction and easy to use.We hope to draw inferences about other cases from one instance and could has some reference to the development of the Hehuatang area.

荷花塘片区的街道、巷道与内部廊道——街巷的形态
Streets, Alleys and Internal Corridors in Hehuatang Area - Street Form

施剑波 吴则鸣
Jianbo SHI Zeming WU

街巷的肌理与尺度往往能映射一个城市或街区的特点，阅读这些街巷网络，可以发现城市演变的痕迹，结合当时的社会文化背景，以期形成比较清晰的发展脉络。本文以南京荷花塘片区为例，在广泛的实地考察以及文献研究的基础上，对街巷的高宽比、交叉口数量、空间体验等进行了整体性研究，并纵向对比民国时期的街巷尺度，横向对比国内外的街巷尺度，提出针对其街巷现状的空间更新与活力激发策略。

街巷是支撑一个城市的骨架。作为一个最基本的功能：从一端到达另一端，街巷承载着城市的日常活动，记录着城市的发展演变、兴衰更替。街巷的物质形态是一个城市的缩影，不同的城市有不同的街巷肌理，研究街巷能方便地把握城市的类型学特征。

位于南京老城南的荷花塘片区，历史悠久，较为典型地反映了明清时期南京传统住区和丝织手工业区的特点。随着1949年后大家族家庭的逐渐解体，多户杂居的模式占据主导，在该片区延续至今，导致频繁出现为扩大人均居住面积而私搭乱建的现象，从而造成该片区逆宜居性发展的现状。梳理该片区的街巷，有助于更全面地了解我们生活的城市，更理性地发现问题，为解决策略提供依据。

Texture and scale of the streets can usually reflect the feature of a city or a district. Reading the street network can help us discover the traces of evolution in a city. When the social and cultural background are taken into consideration, we can form a comparatively clear thread of development. This paper, which takes Nanjing Hehuatang area as an example, is based on the method of field investigation and literature research, studies the ratio of height to width, the number of intersections and quality of spatial experience of the streets. We make comparisons in dimensions: in the time level, we compared the current street scale to that of the Republic times; in the spatial level, we compared the street scale here to that of the western countries. According to such researches we propose a series of strategies to update and to stimulate the vitality of the district.

Street is the skeleton of a city. It has a basic function: from one side to the other of the city, streets carry the city daily activities, record the city development, waxed and waned. The material form of the streets is the miniature of a city, different city has different texture of streets, so study on the streets can help us easily grasp the typological features of that city.

The Hehuatang area, which has a long history, is located in the Old South City of Nanjing. It is a typical reflection of Nanjing traditional residential and silk industrial zones in the Ming and Qing dynasties. With the gradual disintegration of the large family after 1949, many households living in one courtyard become the dominant mode in the district, which appears as the over densely extension and illegally build , results in the current statues of unsuitable for living. Reading the streets can help us understand our life in the city more comprehensive, find the problems more rational and provide basis for solving strategies.

1　现状分析研究

作为南京典型的老城南建筑街区，荷花塘的街巷保留了传统的生活氛围，行走于其中，空间的各个组成要素无不向你传达着"这就是荷花塘，这就是我们生活的地方"。

（1）空间体验

"先见后说。孩子们在说话之前先观察与认知。"——约翰·伯格，《看的方式》

中山南路是毗邻荷花塘的最宽的一条道路，承载着城市繁忙的交通运输和地铁线，尽管如此，在有限的非机动车道和建筑外墙之间，店铺外放的桌椅、店铺招牌、招牌下露出的屋檐，点缀出淳朴的生活气息。在繁忙的街道上尚能限定出安逸的生活场所，荷花塘内部的街道更是几乎成了生活性内院。借助树干和矮墙搭接的横杆，解决衣物的晾晒；曲折的街道减缓了自行车的速度，更适合步行；街道局部放宽成为休憩的场所；铺地的不同划分，更进一步区分出通过和停留的空间。然而，生活性过分突出：私密衣物暴露于公共场所，部分衣物直接晾晒于输电线上，对于空间的美观和体验以及安全性也会造成不利的影响。

1　Current situation analysis

As a typical building block of the Old South City in Nanjing, the streets in the Hehuatang area retain the traditional atmosphere of life, when you walk into the district, every element conveys to you the same message: "this is the Hehuatang, this is where we live."

(1) Space experience

"Seeing comes before words. The child looks and recognizes before it can speak." —John Berger, *Ways of Seeing*

Zhongshan South road, which carries heavy traffic transportation and city subway line in the cite, is the widest road adjacent to the Hehuatang area, however, there are furniture outside the shops, the shop signs, the eaves under the shop signs on the limited pavements which decorated the simple city life. Even in a busy street can define a comfortable place to live. The street inside the Hehuatang area almost becomes a life of the inner courtyard: Residents dry clothes with the help of trunks and parapet laps; winding streets slow down the speed of the bicycles and let it more suitable for walking; some parts of the streets are enlarged and become squares for relaxation and entertainment ; different paving textures distinguish the stay and through space. However, the disadvantages are also obvious, life becomes too conspicuous: underwear are exposed to public places, some clothes are hanged directly on the electric wire, which can damage the spatial experience and cause potential risk.

次一级的巷道多为建筑山墙面之间的狭小空间，少有建筑入口，满足分户、交通、采光需要。部分巷道一侧的建筑底层内凹退让，作为停车和入户空间。这类巷道大多位于背阳面，空间较为压迫和灰暗，使人只想快速通过，难有停留的安逸感。

再下一级的建筑间的内部通廊，多为半公共的灰空间廊道，解决纵向多户的入户需要。这些廊道，有的空间极致压缩，仅作为纯粹的过道；有的略微放宽，置入厨房、储物等辅助功能，是通廊两侧住户共用的，成为入户和交流的场所；有的甚至外化为公共空间，两侧墙壁作为社区公告栏，更像是一个有顶的街道了，这也为邻里之间增加了一个遇见的场所。

（2）街巷尺度

中山南路是荷花塘周边最宽的一条道路，也是城市的主干道。道路的宽度远超两侧建筑的高度，空间较为疏离空旷，因此道路两侧丰富的生活性元素试图在步行通过的尺度上创造一个宜人的空间氛围。陈家牌坊是荷花塘片区靠近城墙的一条街道，街道宽度与两侧建筑物高度相当。相比于其他街巷，它具有一定的公共性和开放性，整条街道有收有放，有通过有停留，且住户和入口数量较多，即使没有商业功能的置入，一样具有活力。片区内部的巷道，其宽度远小于两侧建筑的高度，且入口和窗户数量很少，甚至没有开窗。两侧建筑给人以相互背离的感受，不是很适合行走。

The secondary roadways are mainly narrow spaces between the gables, in these alleys there are little building entrance and just to meet the basic functions of the household, transportation, lighting etc. Part of the roadway opens the ground floor as a entrance or parking space. This kind of roadway are mostly located in the shaded-side, space is more oppressive and gloomy and lack of comfortable space for staying, which makes people want to pass quickly.

The next level of roadways are interior corridors, most are the semi public gray space corridors, they are mainly aimed to solve the household function. The corridors, some are extremely compressed space which only work as a pure passageway; some are slightly broadened to put into kitchen, storage and other auxiliary functions, and become a common space for both sides of the householders to entrance and communicate; some even become the public space, both sides of the wall are used to fix the community bulletin board, which makes it more like a street with a roof, adding a meeting place for the neighborhood.

(2) Street scale

Zhongshan South road is the widest road adjacent to the Hehuatang area. It is also the main road of the city. The width of the road is far more than the height of the buildings in both sides, the space is alienated and open, so we try to create a pleasant spatial atmosphere through the rich elements of life and the comfortable scale of walk on the pavement. Chen Memorial Gateway is a street near the city wall in the Hehuatang area. The height and width of the street are almost equivalent. Compared to other streets, it is public and open, the pass and stay spaces are allocated in a suitable proportion, there are large number of the residents and entrance, even if no commercial function is inserted, it is still vital and comfortable. As for the roadway inside the block, its width is much smaller than the height of the building on both sides. There are few entrances and windows, the building on both sides give people a feeling of departure, it's not very suitable for walking.

建筑之间或者建筑内部的灰空间廊道，有几种不同的尺度。第一种是集约型的过道，1.2～1.5 m宽，纵深较长，两端有采光，空间较暗；第二种是有复合功能的通廊，宽2.4 m左右，将室内的部分功能外置，也为邻里共享提供了可能性；第三种是完全公共化的灰空间过街，高宽比大致为1，在这里，周边住户完全放弃了对这片空间的使用权，连围合空间的墙壁也成为社区公告栏。

There are several different scales in the gray space corridor between or inside the buildings. The first is the intensive aisle which is 1.2~1.5 meters wide, it's long and has light in both ends, the space feeling is dim and dark; the second is a composite function corridor which is about 2.4 meters wide, it puts part of the internal functions outdoor and provides the possibility for neighborhoods to share; the third is completely public gray space crossing street, its height-width ratio is roughly 1:1, the surrounding residents here abandoned the right to use this piece of space, even the walls which enclosed the space have become community bulletin board.

图1 荷花塘地区街道、巷道与内部廊道照片
Figure 1 Photos of streets, alleys and internal corridors in Hehuatang area

（3）整体形态网络

漫步于传统街区，确实能体验到不同的街巷尺度营造的不同的空间氛围及其承载的生活场景，但是对整个片区街巷分布的宏观把控也是十分重要的，因此我们叠合了各尺度等级的街巷廊道。通过分析，我们可以清晰地发现该片区不同层级的街巷形成了较为严密规整的网络系统，这也合理解释了穿行于其中，四通八达的实际感受。

2　纵横向比较

（1）道路交叉口数量横向比较

为了更清晰直观地了解荷花塘片区的街巷密度，我们统计出整个片区的道路交叉口数量，并横向对比同时期国内外的典型城市平面。

可以发现，道路交叉口数量与威尼斯相当，但街区数量却远小于威尼斯。可见，荷花塘的街区尺度偏大，促使居民自发性地在街区内部增加街巷，形成了诸如狭窄的巷子、有顶的通廊、穿越建筑内部的通廊等丰富的街巷形式。有系统地重新组织这些街巷关系、重新划分街区的大小，或许可以在保证步行方便通达的基础上，营造更舒适的居住环境。

（2）纵向比较

在尝试探究形成现状街巷网络的原因的过程中，我们对比了民国时期荷花塘的街巷肌理，查阅了相关的社会背景。我们发现民国时期，该片区仍有不少大户人家在此居住，街区格局多为独门独户的大院落。1949年以后，南京城市中心不断北移，新街口地区逐渐成为城市新的商业中心，城南地区日渐衰退，逐渐演化为城市边缘地带。

(3) The morphology of the overall network

Walking in the traditional street, we can experience the different street scale which create different spatial atmosphere bears and different life scenes. Comprehension in the macro level of the streets distribution throughout the area is also very important, so we all overlapped the street of different scales. Through the analysis, we can clearly find the streets of different levels in the area form a tightly structured network system, it also gives a reasonable explanation why we feel accessible when walking through the streets.

2　Spatial level and time level comparison

(1) Spatial level comparison on the number of crossroads

In order to understand the streets density in the Hehuatang area more clearly and intuitively, we collect the statistics of street intersections throughout the area and compare with the typical domestic and foreign city plane at the same time.

It can be found that the number of street intersections is equivalent to that in Venice, but the number of streets is far less than that in Venice. Obviously, the street scale of the Hehuatang is comparatively large, which stimulates the residents to increase street in the neighborhood spontaneously, such as the formation of a narrow alley, a corridor with a roof, a internal corridor through a building and other streets form. Reorganization of these streets and rezoning of the block size systematically perhaps can create a more comfortable living environment on the basic of guaranteeing a convenient access.

(2) Time level comparison

In an attempt to explore the formation reasons of the current street network status, we compared the Hehuatang street texture in the period of the Republic times, after reading the material of relevant social background, we found that the area still has many large families living in the district in the period of the Republic times, which performed as the single large courtyard pattern. After 1949, the center of Nanjing moved continuously northward, Xinjiekou area has become the commercial center of the new city, southern region is declining and gradually evolved into the city fringe.

1956年开始的房产社会主义改造打破了原先一户一家的居住格局，房屋财产收归国有，再统一分配给单位职工，形成多户聚居的大杂院格局。随着家庭人口的增多，每户为了争取更多的空间面积，不可避免地进行私搭私建，导致整体房屋格局受到破坏，街巷尺度失衡。1966年开始的"文化大革命"则彻底导致房屋产权发生转移，并且多数过激行为使得部分历史久远的住宅遭到破坏，沉淀着历史记忆的空间载体不复存在。

Since 1956, the socialist transformation of real estate movement has broken the previous one large family living pattern, housing properties are nationalized and reallocated unified by the government to the workers, which forms the multi family community courtyard pattern. With the increase of family size, householders inevitably build up illegal buildings in order to get more space, resulting in the destruction of the overall pattern and imbalance of the street scale. In 1966 the "Cultural Revolution" started and led to a thorough transfer of the housing property, some destructive behaviors damaged the historic preservation architectures, the spatial carrier of historical memory does not exist.

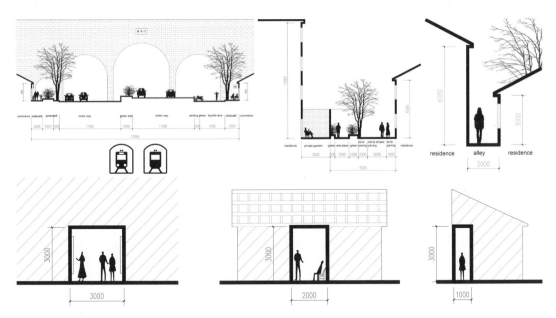

图2 荷花塘地区街道、巷道与内部廊道的剖面
Figure 2 Section drawing of streets, alleys and internal corridors in Hehuatang area

 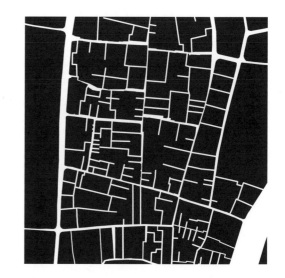

图3 1929年与2016年的荷花塘片区街巷巷道与内部廊道平面
Figure 3 Site plan of streets, alleys and internal corridors in Hehuatang area in 1929 and 2016

可以说，这种逆宜居性的发展趋势是有其历史缘由的，这也造成了荷花塘低层高密度、人口拥挤、人均居住面积小、街巷空间狭窄甚至部分街巷不通的现状。梳理街巷关系、恢复街巷尺度、清除部分私搭私建，可能会还给居民一个更好的居住环境，也是老城复兴的一个契机。

3 结论

类型学作为一种工具，有效地帮助我们梳理了荷花塘的街巷，既直观清晰地展现了街巷空间品质的巨大差异，也反映了当地民众居住现状的恶劣：日照少，通风差，人均居住面积少，交通可达性差。在这里，旧建筑的文化价值的宝贵与居住质量的恶劣再一次发生冲突，居民对于提高居住质量的诉求与建筑师对于保留旧建筑文化内涵的情怀也难以协调。在此条件下，我们针对荷花塘片区的街巷现状，提出如下可能的改善策略：

It can be said that this inverse development trend that the area becomes not suitable for living has its historical reasons, this caused the high density and low height, crowded population, small living area per person, narrow street space and some part of the street is blocked. Comprehension of the streets, restoration of the street scale and clearance of part of illegal constructions may give residents a better living environment, it's also an opportunity for the revival of the old city.

3 Conclusion

Typology, as a study tool, helps us read the streets in the Hehuatang area effectively. It can not only show the great difference in spatial qualities between the streets visually and clearly, but also reflect the bad living condition of the local people: little sunshine, poor ventilation, less living area per person and poor accessibility. The cultural value of the old buildings conflicts with the quality of living, appeals to improve the quality of living from the residents and attempts to retain cultural connotation of the old buildings from the architects are difficult to coordinate. According to the present situation of the Hehuatang area, we put forward the following possible improvement strategies:

①重新组织街巷关系,保证步行的通达性;
②街巷高宽比宜控制在 0.8～1.2m,使街巷既有围合感,又不显得疏远;
③街道可以增加部分放大节点,提供更多公共活动空间;
④部分内部通廊可适当取消,以减少对内部居民私密生活的干扰;
⑤沿街的界面可以考虑置入部分特色手工商业,活跃生活氛围,也是对南京传统手工技艺的继承与发扬。

① rearrange the relationship between the streets to ensure the accessibility of walking;
② The height-width ratio should be controlled in 0.8~1.2m to make the street has a sense of enclosure and not appear estranged;
③ Increase some enlarged nodes in streets to provide more public space;
④ Some part of the internal corridor can be cancelled to reduce the interference of the residents' private life;
⑤ The interface along the street can be considered as manual business function to active the life atmosphere, it can also inherit and develop the traditional crafts of Nanjing.

参考文献

1　LYNCH K. The Image of the City. Cambridge: MIT Press, 1960.
2　MUMFORD L. The City in History: Its Origins, Its Transformations, and Its Prospects. New York: Harcourt, Brace and World, 1961.
3　ROSSI A. The Architecture of the City. Cambridge: MIT Press, 1982.
4　GEHL J. Life Between Buildings – Using Public Space. Van Nostrand Reinhold Company Inc, 1987.
5　卢原义信. 街道的美学. 尹培桐,译. 武汉: 华中理工大学出版社,1989.
6　JACOBS A. Great Streets. Cambridge: MIT Press, 1995.
7　华琳, 王承慧. 南京荷花塘历史文化街区社区特征及发展建议 // 规划60年:成就与挑战——2016中国城市规划年会论文集(17住房建设规划), 2016.

南京城市更新案例研究——以老门东和荷花塘为例
Urban Renewal Case Studies in Nanjing - Two Researches of Laomendong and Hehuatang Areas

韩珂 刘巧
Ke HAN　Qiao LIU

南京拥有两千多年的悠久历史，曾是许多中国古代朝代的首都。虽然历经激烈的社会变迁，南京仍然遗留下了众多历史建筑。这些历史建筑不仅是南京的共同记忆，更是用于历史研究和建筑研究的宝贵财富。保持和维护历史建筑以及传统文化的呼声开始于1990年代。南京的老城南，则是南京城最具代表性的传统文化历史聚集地。这里保存了众多历史建筑，同时也流传了许多传统民间习俗和文化。但是，这一地区的许多建筑和配套设施都太过老旧，不再适于居民日常居住。这一地区的经济同时也需要商业区和旅游景点的推动。因此，整修和重建是这一地区急需的。但问题在于，当城市重建的进程推进到这一区域时，这一区域的历史氛围和传统文化也需要被保留。本文研究了南京老城南的两个重建案例，即老门东和荷花塘重建项目。案例的优点和缺点都得到了研究分析。本文中的分析结果可以对未来的历史建筑和街区重建项目提供一定参考。

1　老门东的城市更新方式

老门东位于古南京的南部，古时候那里集聚了许多商人、贵族家庭和精英家庭，构成了商业和手工艺中心。但现在，老门东却同南京南部其他区域一样面临同样的问题。通过公众意见的汇集，一场城市区域经济发展与传统文化保存的较量逐渐展开，吸引了公众的关注。

Nanjing, which has a history of over 2 000 years, used to be the capital city of many Chinese ancient dynasties. Despite the severe society changings, many historic buildings have survived and still remain in modern times. Historic buildings are not only important parts of the shared memories in Nanjing, but also are of priceless values for historical research and architectural research. The requirements for protecting and remaining the historic buildings and traditional cultures started to be heard since the middle 1990s. The Old South City of Nanjing is the most representative area for traditional Nanjing culture and history. Many historic buildings locate in this area, where lots of traditional folk customs also exist. However, some buildings and auxiliary facilities in this area are too old and not convenient to use. The economy in this area also needs to be promoted by developing commercial district or tourist attractions. Therefore, renewal and reconstruction is needed in this area. But the problem is, when urban renewal process comes to this Old South City, its historical ambience and traditional folk customs also wish to be remained. In this paper, two renewal cases in the Old South City of Nanjing, i.e., Laomendong and Hehuatang renewal cases, are studied. Both advantages and disadvantages of these cases are analyzed. The analysis and results in this paper can be used as a reference for the historic buildings' and blocks' renewal in the future.

1　The renewal ways in Laomendong

Laomendong is an area located within the old south part of Nanjing. The blocks in Laomendong used to gather business people, literary talents and great families in Nanjing at ancient times. The center of commerce and handcraft also lies in Laomendong. But in modern times, Laomendong faces similar contradictions as other areas in the old south part of Nanjing. Moreover, the Laomendong area has always been supervised by public opinions. Therefore, the game between promoting district economy and remaining the traditional culture in this area has drawn many people's attention.

图1 老门东鸟瞰图
Figure 1 Overview of Laomendong

同时，在老门东发展的其他方面也存在诸如此类的矛盾，如是公众集资建设还是私人企业援助，建筑的高度和密度等因素可能阻碍经济的发展。如此种种，老门东的更新开辟出一条政府主导的小尺度、稳步渐进的更新路径。更新结果达到预期，但一些区域还可以进一步开发。下文分析老门东的更新模式。

Contradictions also exist in other aspects of developing in Laomendong. Such as that the funds should be from the public finance or the private enterprise funds, and that the height and the density of the building in Laomendong is restricted which leads to unbalanced future income. With these multiple context, the renewal of Laomendong carries on in a path of small-scale and progressive transforming with government predominance. The renewal results have met the anticipated. But certain parts also can be further improved. In the following of this part, the renewal pattern of Laomendong is analyzed.

老门东的整体更新没有打破旧时的风貌，依然保存了建筑的特色。改造完成后，建筑大部分的功能都已被置换成商业，以副食、小吃街文化、休闲娱乐为主。可以说老门东依旧是南京老城市民俗风貌的存放点。此外，老门东的更新完全尊重了周围重要的文化遗迹，如明长城、内秦淮河。老门东所折射出的文化正是追根溯源的表现，并成功避免了文化断裂和传统商业的衰退。

同时，观察街区和街道，街区和街道的红线也依然保留。建筑物的格局及其组织方式与以前的风格非常相似。只有几条街道、车道和码头的尺寸略有变化，以使交通更加顺畅。建筑物和街道的空间符号沉淀下来的文化仍然保留了南京老城南的味道，而居住和商业经验大大改善。

在微观层面上，通过修复原始的街区和建筑物，形成了包括现代和历史风格的文化象征。大玻璃窗融入典型的传统建筑中，满足现代功能需求，同时体现历史文化特色。更重要的是，为了解决缺乏公共设施的问题，老门东新建了电影院和博物馆，这些设施已经创造出新的商业价值。同时，传统民俗气氛与这些新设施之间的冲突得到了最大的改善。

然而，政府主导的更新也有其缺点。老门东的复兴进程带来了土地所有权的大量变化。以前由房主拥有的住宅和土地现在由开发商开发。随着这些变化，问题和争论也越来越多，政府更关注南京的商业发展开发，而忽视原住民不同的个人需求，选择快速摧毁然后更新重建的方式。

The overall renewal of Laomendong do not break the traditional ambience and follow the architectural features in this area. After the renewal, the residence function of Laomendong has mostly turned into commerce function, in the whole block what is served is mainly mid or low-class snacks, entertainment, and retails. In another word the folk ambience in the Old South City of Nanjing still remain in Laomendong. Moreover, the renewal of Laomendong fully integrates the nearby historic heritages such as the Ming Great Wall and the inner Qinghuai river. The culture reflected by the renewed Laomendong falls back to its historic origin. The fragmentation of culture and fading of traditional customs has been successfully avoided.

Meanwhile, from the view of the blocks and streets, the boundary lines of the blocks and streets also gets remained. The buildings' pattern and how they are organized stay quite similar to their previous style. Only the size of several streets, lanes and yards have been slightly changed to make transportation smoother. The culture experienced from the spatial symbols of the buildings and streets still stay in the sense of the Old South City of Nanjing with residence and commerce experiences greatly improved.

In the micro level, through the modification of the original blocks and buildings, a cultural symbol consisting both modern and historic styles has been created. Large glass windows merges into classic tradition buildings, which satisfies the modern functional needs and reflects the historic cultural characteristics at the same time. More importantly, to settle the problem of the lack of public facilities, cinemas and museums have been newly built in Laomendong. New commercial values have been created with these new facilities. Meanwhile, the conflicts between the traditional folk atmosphere and these new facilities have been avoided with the most effort.

However, the government predominance renewal also has its disadvantages. The renewal process in Laomendong carries on with a large amount of changes of the land's ownership. Buildings and lands used to be owned by the residences now are the properties of the developers. Problems and arguments have risen with these changes. Ignoring the different individual needs of aboriginal residence, the government pay more attention to commercial developments in Nanjing and choose a way of rapid destroying, then reconstructing.

此外，一些仿古建筑只是初看上去像传统样式，但在细节上存在许多缺陷，与真正的历史建筑有很大差异。人们都知道这些建筑是假古董，政府则希望这样的地方成为扮演南京传统的碎片。然而，新建筑物与过去的建造在结构上没有联系，墙壁上的新材料、闪亮的白色涂料和屋顶上的现代瓷砖无法与年代产生关联。毫无疑问，这些假古董不能承载城市的记忆，只是一个现代商业区。在某种程度上，这种再生模式不是考虑古建筑保护和建筑美学的好办法。

老门东的更新模式只是一个南京的典例，实际上在中国很多地方都可以发现这种模式，如武汉花楼街。尽管有明显的效果，但由于国内生产总值快速增长的需求，近20年来，这种再生模式几乎遍布中国的大城市中。最近人们开始评估另一种再生模式，对荷花塘地区的研究如下。

2 荷花塘历史街区改造规划

荷花塘作为明清时期传统居住院落的遗存地之一，是南京历史格局保存最完整、建筑遗存最丰富的街区，其与老门东一起成为南京老城南历史文化保护地段，自2012年以来南京政府开始了改造更新规划。荷花塘更新保护的模式与老门东基本相同，都采用小规模、渐进式的还原改造方式。规划的目的是恢复整体历史格局，修复重点保护建筑。另外，为满足旅游需求，提升商业价值，将一些建筑用做文化展示。

荷花塘的再生与老门东不同，历史上，老门东依托秦淮河成为贸易中心，而荷花塘区域主要为居民居住地。这种差异对于两者再生的打造是有借鉴意义的。

Moreover, some of the renewed buildings only seem to be in traditional forms at the first look. Details in these buildings have many flaws and differ greatly from the real historic ones. People are aware that these buildings are fake classic buildings. Governments want this place playing a role of traditional Nanjing pieces. However the new buildings have no connections to their past plan arrangements or structure, also, the shining white dope on the wall and the modern tiles on the rooftop bears nothing connecting to aging. With no doubts these fake ones can't be a place which bears the city memories but are totally modern business areas. To some degree, this regeneration pattern is not a good way considering ancient architecture protection and architectonic aesthetic.

The regeneration pattern in Laomendong is just a typical case in Nanjing, however this kind of pattern can be found in many places in China, like Hualoujie in Wuhan. Despite of the obvious negative results, this regeneration pattern is implemented almost everywhere in Chinese big cities in the past 20 years thanks to the requirements of speedy increasing GDP. Recently people begin to evaluate another regeneration patterns and the study on Hehuatang area is as follows.

2 Regeneration project in existing Hehuatang historical area

As one of traditional courtyard-housing-style heritage in Ming and Qing dynasties, Hehuatang area is the one that the most well-protected historical patterns and the richest historical buildings in Nanjing. This area was protected as historic and cultural preservation area in Nanjing Old South City, so the government has planned to regenerate this area since 2012. The approach of renovation in Hehuatang area, which is a kind of small-scale and incrementally renewal, is almost the same as in Laomendong. The project aims to restore the whole historical patterns and repair the important protected buildings. Furthermore, some of the buildings will be used as cultural exhibition in order to appeal tourists and improve the economic value.

What is different from the regeneration in Laomendong is that in history, Hehuatang area was the main residential district while Laomendong was one of the business centers depending on the Qinhuai river. That is meaningful in the consideration of the regeneration in both area.

图 2 荷花塘街区现状照片
Figure 2 Existing situation of Hehuatang area

荷花塘规划有两个目标，一是加强对整体历史环境和传统文脉的恢复与延续，政府希望规划能反映出老城南明清时期典型传统民居的特征；二是通过重新利用局部建筑，挖掘经济价值，使其成为老城南今后发展的文化休闲旅游示范点。具体反映为三点：

区域层面上，整体修复并延续典型明清民居特征的格局。建筑进行更新和修缮，完整保留建筑结构和立面形态。一些重要的历史遗存已严重破损，亟须采用地方样式进行重建。此外，内部街巷的传统空间尺度也会保留、还原，街区内的古井、古树等都作为遗迹保存，且控制用做展示和商业空间的更新建筑的形式与材料（如明清传统院落形式、黑瓦灰砖材料特征、建筑高度等）。

The Hehuatang project has two goals. One is that strengthens the restoration and continuance of the whole historical environment and traditional patterns. The government hopes that this project would reflect the classic traditional residential features in the Old South City in Ming and Qing dynasties. The other one is that make the Old South City as an demonstration site of cultural and leisure tourism by resetting the uses of partial historical buildings and tapping economic potential. More details are designed in three parts as followed:

In regional level, the project repairs and continues the typical patterns of Ming and Qing dynasties residential characteristics. The buildings are going to be renovated and repaired by containing the whole structure and the facade style. A few of these important historical preservation are badly destroyed and need to be rebuilt according to their regional style. Besides, the traditional space scale of inner streets are retained and restored. The old trees and old wells in the area are protected as parts of heritage. Those renovated buildings which are used as display or commercial space are also controlled with form and material (the courtyard style in Ming and Qing dynasties, the characteristics of black tile roof and grey bricks, the height of the buildings, etc.).

大部分现状的居住建筑得以保留，另外，政府将改善街区的环境品质，完善居住配套设施以满足当地居民现代生活需求。一方面在街巷内分散地加入小尺度公共空间与绿化，结合明城墙打造市民休闲花园；另一方面增加服务设施，如停车场、公共卫生间、垃圾回收点等。

旅游商业价值的提升和当地经济发展的促动，是改变重点保护建筑利用模式的主要方面。重点保护建筑被规划为民俗体验展馆或者文化博物馆，以展示它们的历史文化地位。荷花塘地区整体历史文化氛围被塑造，从而带动旅游发展，提升经济效益。

3 荷花塘历史街区规划的借鉴意义

自1990年代以来，南京老城南地区就不断经历毁坏拆建、保护整治的改造过程。在多年的经济物质发展、政府建设、民众监督、学者呼吁保护的多方博弈下，荷花塘、老门东的改造模式也是整个老城南改造方式的试验田。当前，以提升传统历史区域整体文化价值带动旅游增长、发展地区经济与提高居民生活的方式是因地制宜、多方需求有机结合的结果。

在荷花塘地区现状保护规划中，传统文化更新中仍然有武断的僵化的做法，导致对传统文化简单化、符号化、装饰化的解读，这并不是真正的传统。政府在建筑形态、使用材料、外观和设计方法上提出过许多特别而僵化的规范，重置了休闲商业建筑的新的功能。这就可以解释为什么会有这么多的假古董和仿古建筑，这种建筑削弱了整个历史保护区的真实性和质感。而且，政府逐渐拆除了建于20世纪中叶和1980年代的典型建筑，以修复荷花塘地区最初的历史肌理。不管这种做法是适宜的还是需要质疑，忽视任何时代的遗存都使文化流失，使历史有了遗漏。

Most existing residential buildings are remained. Furthermore, the government will improve the quality of environment in the area and add more basic residential facilities to satisfy modern lifestyle of people. On the one hand, some micro public space and green lands are added separately in the inner streets. The Ming city wall becomes the part of public leisure garden. On the other hand, some service facilities such as parking, public toilets, garbage collection are added.

The improvement of tourism business profit in the area and the promotion of local economic development, is mostly the way changing the uses of important protected buildings. These buildings are planned to be folk experience space or cultural meseum, that shows their historical and cultural status. The whole Hehuatang area will present the traditional and historical atmosphere of culture, which attracts more tourists and enhancs the local economy.

3 The significant meaning in the existing Hehuatang area project

Since 1990s, the old south city in Nanjing keeps suffering the cyclic process of damage, construction, reconstruction and protection. After so many years of combat between the development of economy, the government requirement, the supervision of citizens and the scholars' appeal for traditional protection, the regeneration mode of Laomendong or Hehuatang becomes the important trial while the proper mode extends to the whole Nanjing old south city. Currently, the mode, that develops the tourism business, enhances the local economy and improves the quality of citizens' living conditions, in the way of enhancing overall cultural value of the traditional history patterns in these area, is the result of suiting the measures to the local conditions and the result of organic combination with many peoples' requirement.

However, in the project of the existing Hehuatang area, there are still some assertive and rigid methods in restoring traditional cultural blocks, which lead to one kind of simplified, symbolic and decorative traditional culture, not the real tradition. The government standardizes too much specific and rigid rules in the type of architecture, the material, the outlook or the design methods while they reset or create the new function in the new leisure and commercial buildings. That may explain why there are many "fake" traditional buildings and pseudo-classic architecture. By the way, this kind of buildings reduce the reality and the quality of whole historical protection area. Besides, the government tends to reduce or remove some typical buildings built in middle 20th century or in 1980s in order to repair the whole original historical patterns in Hehuatang. Whether it is proper or doubtful, it is also the loss of culture or the incompletion of history if ignore any remains in any times.

4 结论

荷花塘街区小规模、渐进式的改造方式会逐渐推广，适应到南京老城的多个历史区段中去。这种在适应城市建设与经济需求的条件下保护城市历史文化、回归城市自身特色的做法仍然需要完善。① 植根于不同地区的历史人文背景，有重点地回归和打造街区的历史特征风貌，避免陷入传统形式同质化现象；② 需要更多地尊重不同历史时期存在的建筑形式与语言，去除更新规划中符号化、形式化的建筑控制；③ 仍然考虑现代和地区生活方式，在历史保护和真实生活空间之间寻求平衡。

4 Conclusion

In the terms of further regeneration of entire Nanjing old city, the approach of renovation in Laomendong and Hehuatang area, which is a kind of small-scale and incrementally renewal, will gradually developed and adopt to other areas in Nanjing traditional old city. The approach that protect the city history and culture by renovating the traditional characteristics of city along with the urban construction and economy conditions, still needs to improve. ① Reconstructing and renovating the historical patterns according to the different traditional background of different area. Do not make the renewal mode homogenous. ② It needs more respect to the characteristics of architecture in different times in the regenerated area. Besides, avoid the rigid and assertive control in project planning. ③ Taking the consideration of modern and local living conditions. There should be balance between the historical preservation and the real living space.

参考文献

1 荷花塘历史文化街区保护规划，南京市规划局，2015年及2013年版.
2 吴晓庆，张京祥. 从新天地到老门东——城市更新中历史文化价值的异化与回归. 现代城市研究, 2015(3): 86-92.

图片注释

1 南京市规划局 [EB/OL]. http://www.njghj.gov.cn/PortalIPS/Library/DefaultSite/a725d883-bf28-46d8-99bf-4d719df6f7c4/1195c32f-2a7d-4d27-9727-9cec1a696b21.jpg.
2 拍摄于南京市荷花塘街区，2016/9/27.

07

师生反馈
Reflection

关于设计工作坊的简评
Some Comments about the Design Studio

马可·德诺西欧
Marco TRISCIUOGLIO

设计工作坊结束后，我的学生助理湛洋收集了来自学生的一些留言和信件。这些内容当然是对于这一经验的报告，同时也是对我们共同工作"事后"的思考。

本科第四年的学生顾家铭说她学习了很多关于老城区的历史以及通过地形学和形态学的方式对城市和建筑的理解。她介绍了工作坊的背景内容："第一个讲座的主题是类型与类型学，一个观念的描述与历史；第二个讲座的主题是原型，从传统形式到新的建筑；第三个讲座的主题是模式，城市形态学，从城市分析到城市设计；第四个讲座的主题是建筑密度，一种理解城市的方式。尽管整个课程体系都略偏向理论化，但是我们还是学习到了很多关于如何在南京老城区有效地进行建造的案例。"关于练习，家铭描述了我们的工作方式："我们一直在遵循一个非常有组织的训练计划，帮助我们缓慢地深入到课程中去。前两个练习与场地有关，是关于南京历史的。然后我们做了几个关于类型与形态的练习，它们能够帮助我们更好地解决城市问题，更好地去理解在城市旧城区更新中极具价值的城市密度和类型形式单元的问题以及与旧建筑有关的功能和历史的其他问题。"

最后，她表达了一点批评的意见："虽然课程没有将很多现实问题纳入考量中且结果更偏向于理论化，但是这仍然是一个有趣的尝试：将欧洲阅读城市的方式与古老的中国联系起来，多么浪漫与迷人！"

After the end of the Design Studio, my student assistant Yang ZHAN collected some messages and letters from our students. Those words are, of course, the report of an experience, but also an "ex-post" reflection about the work we have done all together.

Jiaming GU, fourth year undergraduate student, learned "a lot about the history of Nanjing old area and the topography and morphology way of urban and architecture interpreting". She describes the contents of the Design Studio: "the first lecture is about types and typologies, description and history of an idea. The second lecture is prototypes. From traditional type to new architecture, while the third lecture is about Patterns. Urban morphology, from urban analysis to urban design, and the last one about urban density. A way to understand cities. Although the whole project was a bit theoretical, we were given a lot of examples to show how we efficiently construct our work in Nanjing old area". Concerning the exercises, Jiaming keeps on describing a way to work: "we did follow a very organized way which help us slowly go deeper into the project. The first two exercises were about the history of Nanjing, because our site location. Then we did several topographical and morphological practices to help us know better about the way that it works to solve urban problems, also better understand urban density and the typical form units that could be used in our project to renew the old site, with consideration of the function, the history and other issues of the old buildings."

At the end she feels a critical enthusiasm: "although we didn't put many realistic questions into consideration and the result came to be a little theoretical, it was still an interesting attempt: the European way of city reading combining with ancient China, how romantic and fascinating!"

另一位本科四年级的学生吴逸雯则将她的注意力放在了设计与分析方法上:"由于不同的历史与文化背景,城市形态学对于我们中国的学生是一个崭新的话题。当追溯到这些不同是如何形成的时候,它是十分有趣的。我记得邓浩教授曾经说过因为意大利有着自己的战争历史,城市可能在一夜之间就被摧毁。因此,意大利人不得不寻找自己的方式以重建家园。正因为如此,最为快速和高效的城市形态学才能够被建立。邓教授的话真的很打动我,因为这是一个值得我们中国的城市规划学习的非常好的方式。"

在这个意义上,类型学可以成为一个非常真实的发现:"我认为包括我在内的大多数中国建筑学生认为建筑是一个独特的事物。当马可教授讲授类型这个词的时候,它真的震惊到了我,因为我从来没有想到建筑可以被组合在这样一个全新的系统中去。"

"每一堂课都是对建筑全新的理解,每一次马可给我们的练习都极具才华和创造力。也许开始的时候我们不理解,但是当工作完成以后再回头看,我就能够理解马可为什么这么做(特别是关于城市密度的练习,我发现它非常有趣和富有创造力)。"

The other fourth year undergraduate student, Yiwen WU, pays aphology is kind of a brand new topic for our Chinese students because of different history and culture background, it is a very interesting question when we date it back to how this difference formed. I remembered Professor Hao DENG said about his opinions on this, he said it was because Italy had its own history of war and battle, city was destroyed just in one night. Therefore, Italians had to find their own way to rebuild their hometown. That was why Urban morphology shaped, it was the most quick and efficient way. What Professor DENG said really touched me a lot because I think it is a totally brilliant way for our Chinese urban planning to learn from".

In this sense Typology can become a real discovery: "I think most of Chinese architecture students including me thought architecture as unique thing, when Professor Marco talked about word 'type', it really surprised me because I never thought that building can be grouped in a new system."

"Every class was a new understanding on architecture forine: every exercise Marco gave us after class was talented and creative, some I could not understand before, but after all the work had been done and I looked back, I could find out why Marco did this for us (Especially the exercise of urban density, I found it very interesting and creative)."

这个设计工作坊也是一个将两种文化一起讨论的机会，研究生一年级的学生王安安强调了她的观念："作为一个中国的学生，传统的建筑文化就像是空气一样环绕着我们的生活。但是正因为这种现象，我们常常会忽视它并停留在一个狭隘的视野中。但是在过去的几周中，马可教授所展现出的对中国文化的热爱和对中西方庭院的比较使得我们更加全面地了解庭院的重要性。在课程开始之前，我对类型学的了解还停留在罗西这个名字上。但是通过这个课程，使用类型学的方法去设计荷花塘地块使得我们对类型学更加了解。我们都认为这是一个保持传统中国住区和城市特征的新的方式。"

虽然过程很有趣，但是安安仍然建议在开始的时候对工作进行一个简要的概述："当我去画最后一张图的时候，我发现很难将过程与最后的设计成果组成一个完整的链条，因为我之前的工作缺乏目的性。我认为如果在工作坊开始的时候能够有一个全面的了解可能会使一切变得更好。"

硕士一年级的学生刘巧在她的课程反思中回忆起了一个简短的关于庞贝的练习："这个关于庞贝的练习以及一系列的讲座给了我很大的帮助，至少让我能够认识什么样的建筑和模式是同样的类型。这在我们学习城市和城市的发展中是极其重要的。当我第一次接触到'形态学'和'类型学'这两个词时，我很困惑，无法理解它们真正的含义，所以我才加入了这个工作坊，现在我已经学习了很多关于类型学和形态学的知识。"

怀着热情、自信与好奇在整个区域的类型地图中设立了一个目标："我们对南京的城市发展做了大量的研究，我对展现荷花塘的三张图版感到自豪，我更感兴趣的是我们完成了类型学的地图。它看起来更像是我们研究并绘制了荷花塘地区所有的房子。这是非常专业和有意义的。"

The Design Studio is also an opportunity to make two different cultures talk each other, as An'an WANG, first year master student, highlighted in her message: "As a Chinese student, traditional architectural culture is like the air behoving in our life. But precisely because of this, sometimes we may ignore it or stay in a narrow field of vision. But in the past weeks, professor Marco's love of Chinese culture and the comparison of the Chinese and Western courtyard make us understand the significance of the courtyard more comprehensively. Before the start of the course, I know typology only stay in the name of Aldo Rossi. But by the studio, using a typological method to design Hehuatang area lets us become more familiar with typology. We all think that it is a new way to maintain the traditional Chinese settlement or the characteristics of the city."

Even if "the process is very interesting", An'an suggests giving an overview of the work at the beginning: "when I began drawing the last board, I found it is hard to connect the process with the final design results to form of a complete chain of logic, because before I do these practice lacking of destination. I doubt it would be better if we had a full understanding of the last work at the beginning of our studio."

In her notes rethinking about course, Qiao LIU, first year master student, remembers a short exercise on Pompei: "the exercise about reading Pompei and the series of lectures helped me a lot, at least I could recognize what kind of architecture and patterns are same type and what are not. It is basic and really important when we study the city and learn its urban development." "when first I read the words 'morphology' and 'typology', I was confused and can't understand what it really means. So I joined in the studio. And now I think I've learned more about typology and morphology."

With passion, proud and curiosity, I set a goal in the typological map of the entire area to reach: "we did a lot research about the urban development in Nanjing. I was proud of three boards that show what happened in Hehuatang. I am more curious about what if we complete the typological map. What it seems that we research and draw all the houses in Hehuatang area. It must be really professional and meaningful."

刘巧从学生的角度对教学方法给出了自己的理解:"我认为建筑类型学是开始设计一个区域的很好的方法。通常情况下,我们从场地、道路以及场地周边的信息以及轴线等开始,真正关注房屋与场地的关系。这就像是我们从最小的单位(房子)开始,然后进行组合,适应道路、环境以及现状,然后构成整个的形态。这是一个比较合理的方式,因为这才是城市形成的过程,模拟了城市的发展。这与我以前学过的其他课程,可以得出相反的结论。"

另一位学生则真诚地回到了两个文化的问题上:"你所做的工作就是把西方的理论和中国传统的聚落模式联系起来,这是非常有意义和价值的。对我来说,我不会有勇气在英国或法国做同样的工作,因为我对其他国家了解得很少。中国的历史与文化是很厚的一本书,我们所看到的现存古建筑的传统形式和符号都很少直接用在我们现代的设计上,例如传统的窗户、中国的红灯笼等等。但是它们都是区别于其他国家建筑的极为重要的特征。"

他也认识到老师的认知也有个提高:"总的来说,中国的学生与其他国家有很大的不同。在课堂上,我们往往习惯于默默地听老师讲课和记笔记。但是这是一个很好的机会让我们在课堂上更积极地参加讨论。俗话说,教学相长。我们可以互相学习,我们学习新的理论,你也可以了解中国。和你在一起是一段非常愉快的时光,我会永远记住这段经历。"

Qiao LIU gives the best comprehension of teaching methods from the perspective of student: "I think it was a good way to start from architecture typology when we began to design the area. Usually, we start from the location, the roads, the surrounding information about the site, the axis and so on, and really pay attention to the relationship about what exactly the house should be like if putting the house in the area. It seems we start from the minimum unit (house), then how to combine the unit, then adapt to the roads or the environment or the current situation, finally the morphology is formed. This is more reasonable because it is the process of the city formation and we are simulating the city development. While in other design courses I learned before, It can be concluded the opposite process."

Another student comes again back to the question of the two cultures in a very honest way: "the work you have been doing, connecting western theory and Chinese traditional settlement patterns, is very meaningful and valuable. For me, I would not have the courage to do the same work in England or France because I did know less about other countries. The history and culture of China is a very thick book: the traditional forms and symbols that we see from the existing old buildings, such as traditional windows, Chinese red lanterns and so on, are seldom directly used in our contemporary design, but they are of vital importance to distinguish from other countries' buildings."

And he realized that there is always an improvement in the knowledge of the teachers, too: "all in all, Chinese students are very different from other countries'. In class, we are used to listening to the teachers and taking notes silently. It is a great chance for us to be more active and have more discussions during the class. A saying goes that teaching benefits teachers as well as students. We can learn from each other, while we students learn the new theory, you can also know much about China. It is a very pleasant time to be with you, and I would keep this experience in my mind forever."

硕士一年级的学生吴则鸣称这个教学方法是可识别的和系统的。他写道："这个学期开始的时候，我对南京并不了解。但是调查研究的部分帮助我快速了解了城市结构和历史演变，论文阅读和理论学习帮助我了解类型学和形态学。然后经过一些失败的尝试和误解，例如如何留白、如何对待道路和巷道以及如何处理不规则的街区变形，我发现自己学到了很多东西。"

本科四年级的学生朱鹏飞解释说："在课程结束之后，我带我的朋友再次拜访了荷花塘地区，因为它真是一个美丽和历史悠久的地方，值得我们去研究，所以在课堂上首先吸引我的是这个地块本身。坦率地说，这不是我第一次关注中国传统的乡土建筑和对应的生活方式。"

"然而，这一次我们是通过一个非常新颖的基于形态学的研究方法。宏观上，我们整理概括了该区域内建筑物的类型，然后提取了其中包含的空间逻辑。在保留历史空间框架的前提下，用它们来创造新的类型去更新这个美丽而略有脏乱的地区"。他补充说："事实上，建筑业目前正在衰退。相反，更多关于文物保护的工作正在等待我们的关注和投入。有时我们只处理一两个旧建筑，但当涉及一个像荷花塘这样巨大的地块时候，我们需要做全面的考虑。处理这种问题的方法是我在这个课程中最大的收获。"

Zeming WU, first year master student, appreciates the teaching method as "distinguishable and systematic". he wrote: "as I'm not familiar with Nanjing at the beginning of this semester either, the research and survey helped me to comprehend the structure and historical changes quickly. Then the paper reading and theory learning helped me know what is typology and morphology. After some failure attempts and misunderstandings of how to "stamp" the blank, how to treat with the roads and alleys and how to deform while facing an irregular block, I find myself really learnt something."

"After the end of these lessons", explained the fouth year undergraduate student Pengfei ZHU, "I lead some friends of mine to visit Hehuatang area again, because it is really a beautiful and historical place which worth our research. So, what attracted me firstly during the lesson was the area itself. To be frank, it's not the first time that I placed attention to traditional Chinese vernacular architecture and their way of living."

"However, this time we got a pretty fresh method of doing our research based on morphology. From a macro-perspective, we collected types of the buildings inside the area and then extracted the spatial logic contained therein, which can be used to create new type to renew this beautiful but dirty and disorder area with its historical spatial framework remained". And he added: "Actually, construction industry is declining currently, and on the contrary, more work concerning heritage conservation are waiting for our attention and input. Sometimes, we just deal with one or two old buildings. But when it comes to the situation that a huge field like Hehuatang area, we need to do overall consideration. The way of thinking to deal with that is what I found the most useful treasure in this lesson."

后记
Epilogue

基于城市形态类型学的
当代设计教学实验
鲍 莉

A Contemporary Design Teaching Experiment
Based on Urban Morphology Typology

Li BAO

环境意识、人文诉求与科技创新并重是全球社会的发展大势,经历了30多年来人类史上规模最大、速度最快的城镇化进程,中国正在走向深度城镇化的发展。之前注重量的累积的"增量发展"转变为注重质的提升的"存量优化、精细发展"成为必然。国家战略上,这既是现代化的必由之路,也是一项重要的民生工程。

在此背景下,除了千篇一律的寻常高层街区或是所谓延续传统风格的高档封闭住区,中国的历史城市的更新可以呈现出怎样的面貌呢?决策者和设计师或许可以考虑第三种方式:借鉴建筑学经典的城市形态学与建筑类型学理论方法以回应中国当代城市问题,探索基于城市既有的建筑类型与形态去设计新的可持续街区和建筑类型的可能方法路径。

在这样的思路和背景下,东南大学建筑学院与意大利都灵理工大学建筑学院合作设置了联合设计课程,目标是学习以形态学和类型学的方法去阅读现有城市与建筑并探索城市及建筑的更新设计策略。

课程设计选址南京老城南的荷花塘地块,是典型的出于城市风貌保护与棚户区改造两难却需兼顾的地块,也是南京市主城区最大的一块城市更新"难题"地块。借此,重新认识由萨韦里奥·穆勒托里和奇安弗兰科·卡尼吉亚建立、并因阿尔多·罗西而传播至世界的意大利形态学的设计理论和方法工具,进一步论证了形态学作为城市设计工具的有效性,并在对当代中国城市形态研究和城市设计中体现并丰富它的价值。

Environmental awareness, humanistic demands as well as scientific and technological innovation are the development trend of global society. Having experience of more than 30 years of urbanization with the largest scale and the fastest speed in the human history, China is moving towards the urbanization to the deepest extent. The early"incremental development"focusing on the" incremental development" is transferring into the"stock optimization, fine development"stressing the quality enhancement, which is the only way to modernization and an important livelihood project as a national strategy.

Which could be, nowadays in China, the outcomes of an urban re-generation project? Should we really decide between ordinary districts made by tall anonymous buildings or gated communities closed into inner villages dominated by"fake"restorations in Chinese traditional style? There might be a third way to designers and decision-makers: to take architecture classic urban morphological and architectural typological theories as in instrument to response the problems emerging in China's contemporary cities, and to explore the new approach to design urban quarters and buildings sustainably upon the existing typological and morphological essence of the city.

Along with such thinking and national context, the School of Architecture at Southeast University cooperating with the School of Architecture at Polytechnic University of Turin has set up a joint design course, which is to attempt to read the existing cities and buildings in morphological and typological approaches and explore the regeneration design strategies for cities and buildings.

The site of design named"the Hehuatang plot"in the Old South City of Nanjing, is a typical plot in historical area which is a dilemma but have to balance the townscape conservation and the renovation of shantytowns. It is and the remained largest"thorny"site for urban regeneration among the major urban districts of Nanjing. Taking advantage of this case, the design theory and methodological tools of the Italian morphology theory established by Saverio MURATORI and Gianfranco CANIGGIA and spread to the world by Aldo ROSSI were taken and re-understood. The validity of morphology as an urban design tool has been further verified, and its values have embodied and enriched in the morphological researches and urban designs of the contemporary China's cities.

课程于2016年9—11月间展开，由意大利都灵理工大学马可·德诺西欧教授和东南大学建筑学院鲍莉副教授联合设题，吴锦绣老师参与授课，来自两校的姜蕾博士和湛洋硕士担任助教。这是一次中西合作、本硕贯通的联合设计实验，参与学生来自建筑学本科四年级和研究生一年级。

整个教学过程包括几个循序渐进的阶段。首先，学生需要从地理、历史和社会学的角度对南京及城南地区以及荷花塘地块进行整体调研分析；其次，在混杂的城市肌理中辨析若干典型的空间组织类型：院落式住宅、狭窄的城市巷道、被穿越的院落/建筑和临街的建筑并抽象出各自的原型；之后，学生需要通过草图、模型和方案图等方式去剖析阅读这几种类型原型并依据场地定位、功能属性、生活方式等设计与策划要求衍化出新的类型，再对其进行归置、组织并调整以生成新的建筑集群模式；最后，运用这些集群模式创造出地块内既源于又有别于原有传统城市形态的新的城市肌理。这其中，我们假设只保留场地内的文保类及历史建筑，其余部分的住区宅院均可纳入重新设计的范围。

学生被分为本硕混编的四个大组，分别从历史、现状、愿景等方面找到切入点展开分析与设计。每组先是合作完成城市形态分析、建筑类型剖析与原型提取，并共同确定地块更新策略及完成总图结构。在此基础上，每位同学再继续深化完成基于原型的建筑类型衍生、空间节点及单元居住院落设计。研究生同学并继续深化到单体建构层面的设计，从而完成从城市设计1:500到单体详图1:20的大跨度多尺度的设计推进。

The studio was carried on from September to November 2016. The topic was determined by Marco TRISCIUOGLIO(Politecnico di Torino, Italy) together with Professor Li BAO. Professor Jinxiu WU took part in it and the assistants were Dr. Lei JIANG and Master Yang ZHAN. This is a teaching experiment of cooperation between China and the West and is devoted to a mixed team of students (attending the fourth year of the Bachelor Program and the first year of the Master Program in Architecture)

The whole program has consisted in a couple of gradual stages as following. Firstly, the students were asked to investigate and study generally on the city and historical area in terms of geographical, historical and sociological approaches; then to recognize, in the existing and chaotic urban fabric, some recurrent types of spatial composition: the courtyard-houses, the narrow lanes, the courtyard houses with roof-covered corridors, and the lined building defining the streets as well. Furthermore, with sketches, models and schematic drawings, students needed to elaborate and abstract the corresponding prototypes and develop the new ones upon the analysis of site location, surrounding condition, users and living modes. Finally, these new types would be composited, organized and adopted into new, but rooted in tradition, urban "patterns".In such process, it was possible to imagine that the historical and heritage buildings can be kept and merged into the new urban fabric.

The students were grouped in four teams mixed with undergraduates and postgraduates. Each of them has elaborated the analysis and designs from as historical, current conditional and visional approaches. Each group started withcooperation to complete urban morphology analysis, building type analysis and prototype abstraction, and determined together the urban renewal strategy and the master structure. Moreover, each member would further completed the building type derivation , spatial nodes and courtyard living unit design based on their new prototypes, afterwards the postgraduates further developed the design into construction and detail level. Thus the whole design has been implemented in a large-span and multi-scale from 1:500, urban design to the construction details of 1:20.

由于是本硕贯通课题，两个阶段学生的课题周期分别为8周和12周。研究生前后各多2周时间，完成专为其设置的若干侧重理论与研究的训练，包括关于城市形态学及建筑类型学的文献阅读、古罗马城市庞贝的建筑类型、城市肌理演变分析以及撰写涉及本土建筑类型及其演变的小论文等。本硕贯通的组织方式，也成为教学上一种有益而有趣的尝试。前期工作以本硕混编的小组为单位展开，研究生通过交流汇报及共同工作的方式与本科生分享理论研究成果，开拓整个设计团队的学术视野，也对整个课程的理论背景与现实基础有更整体和更深入的认识。

本次联合课程设计的价值在于对经典设计理论和方法的学习与探新，在面对中国当代城市问题时，尝试运用诞生于1950—1960年代的意大利、德国和英国，但过去几十年则逐渐被遗忘的两个认知与设计城市及建筑的概念工具：形态学和类型学。并希望通过重新发掘、辨析和推进，使之可以成为城市更新与设计的工具，从而在当下的设计中观照城市过去与未来的连接。

本次设计教学实验的完成及成果的结集出版，离不开所有参与课程设计的同学们，他们从懵懂到明悟，从犹疑到投入，认知与学识的收获想必已超越课堂与图景。联合教学项目也得到了东南大学建筑学院和院长韩冬青教授一直以来的大力支持和关心。

付梓之际，还要感谢群岛工作室的秦蕾编辑在排版工作中的建议和付出，更要向东南大学出版社戴丽老师和魏晓平编辑在编辑、出版中的给力支持致以特别的谢忱。

The studio are devoted to a mixed team of students (attending the fourth year of the Bachelor Program and the first year of the Master Program in Architecture). The durations for them are 8 weeks and 12 weeks respectively. To take use of extra4 weeks, the additional exercises were given to the master students for theoretical or analytic training: literature reading about typology and morphology, typological reading exercise of Pompeii's building (and the transformation of the urban fabric in that ancient Roman city), and writing short essays upon their work. As a mixed team with both bachelor and master students, that also provided professors a chance to explore some beneficial and interesting teaching methods. The first phase of the work was based on mixed groups. Through presentations they shared the study reports by master students, together working helped also the bachelor students to better understand the theoretical background and the realistic basis of the whole project, and to broaden their academic horizons.

At the same time, the opportunity was much greater to refresh and enrich two conceptual instruments: morphology and typology, which emerged in the Fifties and Sixties of last century (between Italy, Germany and UK). This teaching experiment attempted to rediscover, critique and refine them to become the instrument to urban regeneration for architects again, so that to reflect the urban permutation linking the past and the future at present.

The completion of this design experiment and the publication would not succeed without any of the students involved in the design studio, who were inspired and dedicated, and who should have learned more and gain more beyond the classroom and imagination. Also, this project as usual gained abundant support and concern of the School of Architecture and the Dean, Prof. Dongqing HAN.

Standing upon the threshold of publication, my thanks also go to editor Lei QIN for her excellent suggestions and contributions, and my special gratitude to Li DAI and Xiaoping WEI of Southeast University Press for their great support in editing and publishing.

谨以此书向东南大学建筑学院 90 周年院庆献礼！
This book is to celebrate the 90th anniversary of the School of Architecture of Southeast University.

图书在版编目（CIP）数据

类型的恒在与城市的蜕变：南京城南荷花塘地块及住区建筑更新设计 / 鲍莉等著. -- 南京：东南大学出版社，2017.11
（东南大学建筑学院国际联合教学丛书 / 张彤主编）
ISBN 978-7-5641-7471-2

Ⅰ.①类… Ⅱ.①鲍… Ⅲ.①建筑设计 – 教学研究 – 高等学校 Ⅳ.① TU2

中国版本图书馆 CIP 数据核字（2017）第 269350 号

类型的恒在与城市的蜕变

责任编辑： 戴　丽　魏晓平
责任印制： 周荣虎
出版发行： 东南大学出版社
社　　址： 南京市四牌楼 2 号（邮编 210096）
网　　址： http://www.seupress.com
出 版 人： 江建中
印　　刷： 上海雅昌艺术印刷有限公司
开　　本： 889mm×1194mm　1／20
印　　张： 9.5
字　　数： 456 千字
版　　次： 2017 年 11 月第 1 版
印　　次： 2017 年 11 月第 1 次印刷
书　　号： ISBN 978-7-5641-7471-2
定　　价： 98.00 元
经　　销： 全国各地新华书店
发行热线： 025-83790519　83791830

版权所有，侵权必究
本社图书若有印装质量问题，请直接与营销部联系。电话（传真）：025-83791830